TQM for Purchasing Management

James F. Cali

McGraw-Hill, Inc.

New York St. Louis San Francisco Auckland Bogotá
Caracas Lisbon London Madrid Mexico Milan
Montreal New Delhi Paris San Juan São Paulo
Singapore Sydney Tokyo Toronto

Library of Congress Cataloging-in-Publication Data

Cali, James F.
    TQM for purchasing management   / James F. Cali.
        p.   cm.
    Includes index.
    ISBN 0-07-009623-6
    1. Industrial procurement—Management.   2. Purchasing—Management.
3. Total quality management.   I. Title.
HD39.5.C35   1992
658.7'2—dc20   92-23123
                                                                    CIP

    2 3 4 5 6 7 8 9 0   DOC/DOC   9 8 7 6 5 4 3

ISBN   0-07-009623-6

*The sponsoring editor for this book was James H. Bessent, Jr., the editing
supervisor was Ann A. Craig, and the production supervisor was
Suzanne W. Babeuf. It was set in Palatino by McGraw-Hill's Professional
Book Group composition unit.*

Printed and bound by R. R. Donnelley & Sons Company.

Contents

Part 2. A Model for Implementing TQM in Purchasing

An Open Letter to Purchasing Managers and Their CEOs

When a company opts for Total Quality Management in its purchasing operation, it embarks on a new and open-ended journey. Behind it lie all the old ways—tried, true, and tired. Ahead lie exploration and discovery, trial and error, challenges and real rewards.

TQM is still too new to be an exact science. (Its underlying principles go back a long way, but it began its wildfire sweep through American industry in its current form barely five years ago.) But TQM is *not* too new to have been tested, expanded on, and fine-tuned by dozens of major U.S. companies—companies that are enjoying the fruits of their labors in the marketplace right now.

One of the main purposes of this book is to pass on the benefits of these corporations' experience (both good and bad) in applying TQM to purchasing operations. Another is to facilitate the discussions between purchasing manager and CEO that are a necessary part of the process. As a twenty-year veteran in the purchasing arena, I can tell you from personal experience that any or all of the following questions are likely to come up in these discussions:

- How can we apply TQM to our supplier strategies?
- What are other companies doing with TQM and suppliers?
- How should we assess and incorporate customer input?
- What are our training needs?
- How do we go about establishing teams?
- What benchmarking—if any—program should we implement?
- What impact can we expect TQM to have on our purchasing operation?
- How can we measure our progress?
- Is the use of TQM methods in our purchasing operation likely to have an effect on our ability to win the Baldrige Award?

To provide answers to these questions, I have drawn on many sources, including my own experience using TQM in purchasing management at Westinghouse, ITT, and Harvard Industries. Throughout the book, an effort has been made to provide a full range of guidelines, tools, definitions, and suggestions. Each of the major steps involved in implementing TQM is illustrated with real-life applications and the lessons learned by those who have gone before. Benchmarking, a cornerstone of TQM, receives special attention, as does purchasing's role in applying for the Malcolm Baldrige Award.

A final word about the all-important dialogue between the purchasing manager and the CEO: Because TQM is a method that requires testing and alteration for a perfect fit, its success at any company depends on the corporate climate—a climate that is inevitably determined from the top down by the CEO. And here it cannot be emphasized too strongly that *nothing is more important, at every stage in the TQM process, than the CEO's blessing and ongoing support*. By providing that support—in the form of encouragement, patience, understanding, and flexibility—the CEO can provide the best assurance that the TQM effort will be a resounding success.

James F. Cali

TQM for Purchasing Management

PART 1

TQM's Impact on Purchasing

1
What Is Total Quality Management?

Introduction

TQM is destined to become one of the most frequently used acronyms of the 1990s. Pick up just about any professional journal these days and you are likely to see some reference to Total Quality Management. Growing numbers of chief executives in the United States and abroad are becoming convinced that TQM principles are the wave of the future. Some executives, misunderstanding TQM's importance, may only pay it lip service. Others, already benefiting from its successful implementation, are now enthusiastic supporters of the TQM approach.

And what, exactly, is the TQM approach? In the briefest possible summary, it is:

- A foundation for continuous improvement
- A philosophy for running a business
- The right way to manage
- Total people-empowerment

- A focus on the customer
- A commitment to quality
- An investment in knowledge

Many of us understand only a small part of TQM. The situation is like the old adage about touching an elephant while blindfolded. If you don't know it's an elephant to start with, you may be misled by whatever part of it—the trunk, ear, tusk, leg, or tail—you happen to be touching. Similarly, people tend to interpret TQM only in terms of the part of the process they're familiar with, not realizing that TQM actually has a wide variety of elements, themes, and principles.

Different levels of an organization often have different understandings of TQM. A factory worker, for example, may believe that TQM involves only the quality of the part he or she is working on, while the CEO interprets TQM as being every aspect of the company's activities including clerical work and product quality. Effective implementation of TQM, however, requires that there be a consensus on its definition and understanding at every level of the organization.

TQM: A New Perspective

Imagining Optimum Results

TQM has generated many new and some not-so-new buzzwords. *Imagineering* is a term that has been around for a while and is often used to define an essential first step of TQM. One of the best explanations of what imagineering is, is illustrated by a story about the famous golfer Jack Nicklaus. During a golf tournament, a news reporter following the golf circuit finally worked up the nerve to approach the famous golfer. After apologizing for the interruption, the reporter questioned Nicklaus about the extended time he spent addressing the ball (preparing to swing). Nicklaus's response was that while he stood over the ball, he imagined the "perfect shot": the position of his body as he started the backswing, the feel of the club as he hit the ball, the trajectory the ball would take, the point at which it would hit the ground, and the sound it would make as it hit the pin. Nicklaus would then take his shot, compare the result with his imagined perfect shot, and make mental corrections for the next time. This was his way of continuing to improve his game.

Similarly, a world leader in custom total quality improvement systems, Dr. A. Gunneson, CEO of the Gunneson Group International Inc., defines TQM by imagining the ideal work-place. Imagine what a company would be like, he suggests, in which every employee had everything needed to do the job right the first time, every time; where all the delays and frustrations most of us experience every day were gone; and where each worker was able to find out what the next person in the cycle needed, all the time, every time. Imagine never having to say, "It's the best I can do under the circumstances." For all the complexities of executing a TQM initiative, this description perfectly captures the results sought.

TQM—A New Paradigm

One approach to understanding TQM is the use of paradigms. A paradigm is basically a set of ideas, usually unwritten, that people have learned through experience and that define the "conventional wisdom" about the rules of nature and life. A paradigm acts as a mental filter. It limits the way we think about things by erecting a set of boundary conditions that are often more perceived than real.

The following are paradigms that seemed right at the time. Hindsight tells us just how wrong these paradigms were.

The ordinary horseless carriage is a luxury for the wealthy; it will never, of course, come into as common use as the bicycle.
—LITERARY DIGEST, IN 1889

I think there is a world market for about five computers.
—THOMAS J. WATSON, CHAIRMAN OF IBM, IN 1943

There is no reason for any individual to have a computer in the home.
—KEN OLSEN, PRESIDENT OF DIGITAL EQUIPMENT CORP., IN 1977

Until very recently, the paradigm used by most business managers went something like this:

- People dislike work; they work only for money.
- Few people are capable of self-direction.
- Management's job is to keep jobs simple and to supervise closely.
- Only when people are closely controlled will they meet established standards.
- If employees are treated firmly but fairly, they will respect the authority of their supervisors.
- To assure quality, the product must be inspected and the nonconforming areas reworked.
- The end result is that the customer will buy what is available.

These kinds of beliefs are what gave rise to the "old" quality culture. The basic concept, of course, with its principles of inspection for conformance, identification of defects, and assignment of blame for any nonconformance, was deeply flawed, because it was focused on measuring the product, not on improving the process.

Using TQM, a new paradigm prevails:

- All activities are based on the belief that the customer comes first.
- All energies are directed at striving for continuous quality improvement.
- An ongoing effort is made to eliminate all waste, with waste defined as the endless cycle of scrapping, reworking, and retesting, rewriting, reanalyzing, and redesigning.
- It is assumed that people *do* want to contribute, that they therefore represent untapped resources, and that ways must be found to harness these resources.
- The ideal working climate is one in which everyone can contribute to continuous improvement.

The new paradigm requires us to think differently about the customer base and develop a partnership with the customer. The new paradigm requires us to change our approach:

- *From* one of correcting defects and nonconformance, *to* one of establishing processes and procedures that will prevent defects
- *From* one of using inspection to "correct" quality into the product, *to* one of using consensus to design and build quality into the product
- *From* one of accepting a level of defect as normal practice, *to* one of establishing a culture of continuous improvement of the process
- *From* a win-lose mentality that involves "beating on" suppliers, *to* a win-win approach that involves trusting and working with suppliers

TQM: A Working Definition

To provide a common reference point, consider Figure 1-1.

The TQM way of running a business should involve a change in thinking from "If it ain't broke, don't fix it," to "If it ain't perfect, continue to improve it." Most of us have been conditioned to only work on the problems, on the negatives. Rarely have I attended a meeting in which the focus has been on improving the things already being done well.

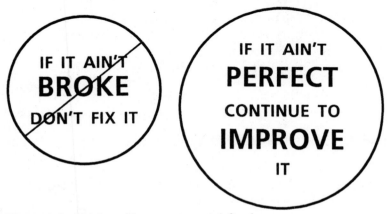

Figure 1-1. Total quality management defined.

The TQM approach involves the following equation:

Improved quality = improved productivity =
improved bottom-line results

Many of us are already accustomed to the belief that productivity leads to bottom-line results. However, the assumption that quality = productivity is a new concept, and some people have difficulty accepting it. But to accept the principles of TQM, one must become a believer in the idea that improved quality = improved productivity.

One Fortune 500 CEO tells the story that he knew his people had reached this realization during a board meeting when the first thing they discussed was quality. The discussion became so lively that it took the entire allotted time. "The numbers" were never discussed.

The Cost of Quality Is a Barometer of TQM

In the world of TQM, another term, Cost of Quality, or COQ, has caused some confusion. Put very simply, COQ identifies areas for process improvement. The term derives from the need to quantify and communicate quality in the language of money—i.e., the language of upper management. Although not new, COQ takes on added importance with the focus on quality. According to one view, COQ is the cost of making a product conform to quality standards. According to another, it is the cost of *not* conforming to quality standards—i.e., waste.

For simplicity's sake, I prefer the following definition:

The cost of quality = the cost of conformance (achieving quality)
plus the cost of nonconformance (waste)

Costs related to quality are usually separated into at least three areas:

1. *Prevention costs.*   These costs are associated with all the activities that focus on preventing defects or nonconformance with

quality standards. In many organizations this includes all the people in the quality department who inspect the product, as well as the operators who are now increasingly serving as their own inspectors. Also included in this group are "supplier capabilities" and the actions that suppliers take to assure conformance. Supplier reviews, supplier ratings, purchase-order technical data reviews, and supplier quality-planning also belong in this group of costs.

2. *Appraisal costs.* These costs are associated with measuring, evaluating, or auditing products to assure conformance with quality standards and performance requirements. Purchased material is categorized in this element and includes purchasing appraisal costs, qualifications of supplier product, and source inspection and control programs.

3. *Failure costs.* These costs are associated with evaluating and either correcting or replacing defective products, components, or materials that do not meet quality standards. Failure costs can be either *internal failure costs* that occur prior to the completion or shipment of a product or the furnishing of a service; or *external failure costs* that occur after a product is shipped or a service is furnished.

The basic relationship among the three types of costs is that dollars invested in prevention and appraisal can substantially reduce failure costs. In addition to reducing expenses, the reduction in external failures results in fewer customer complaints.

A poll taken by the American Society for Quality Control estimated the cost of quality as a percent of revenue. Figure 1-2 shows the results of that poll. The experts estimated that the U.S. percentage of greater than 20 percent is at least double the percentage for Japanese companies.

IBM estimates that their cost of conformance is one-third the cost of nonconformance. (See Figure 1-3.)

The COQ concept is also well known to those in the defense industry. The government has several regulations on quality, including Mil-Q-9858A-Quality Program Requirements. The government performs periodic reviews of their suppliers (contractors) to assure that their procedures reflect the requirements of the Mil

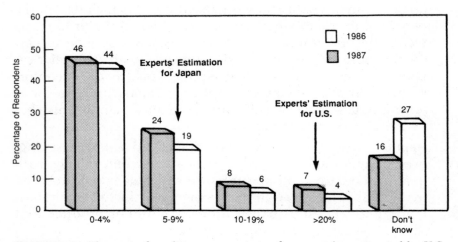

Figure 1-2. The cost of quality as percentage of revenue (as estimated by U.S. executives). (*ASQC/Gallup Poll*)

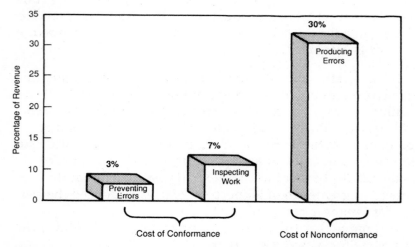

Figure 1-3. The cost of quality, according to calculations by IBM. (*IBM Quality Institute*)

regulations; the contractors comply with their own procedures; and the quality data is being used by the contractor to implement positive corrective action. Contractors are also responsible for assuring that their own suppliers have acceptable programs.

Summary

However broad and all-encompassing it may be as a concept, TQM has very specific applications in the world of business and industry. Its focus on the customer and on constantly refining the process and reshaping the working relationships of those responsible for the product make it an ideal tool for achieving industry-wide improvements in the decade to come. Its effect on purchasing has already begun to make itself felt. Its future impact on the field may well be summed up as a focus on quality as the basis for survival.

2
Why Are We Talking About TQM?

Introduction

Managers and executives would probably explain the widespread enthusiasm for TQM in one of the following ways:

, "It's the right thing to do."
"It seems to work."
"Our competitors are doing it."
"We need it in order to compete globally."
"Our customers are doing it."

While each of these reasons has validity, an even stronger driving force behind all the current interest in TQM is the reaction of American companies to the deteriorating position of the United States as a global competitor. The declining U.S. position has been dramatically highlighted by the Japanese increase in market share in automobile and consumer electronics.

The reaction by U.S. firms has been to adopt the principles of Total Quality Management originally taught to the Japanese by Americans and later implemented by Japanese industry. Many American companies have achieved success by refocusing their energy and attention on quality and by making satisfied customers their top priority. In these companies, continuous improvement in quality is emphasized throughout the organization. Indeed, the focus on quality involves satisfying the customer through the use of the entire organization.

But the important lessons learned by American firms from Japanese companies are only the most recent step in an ongoing industrial evolution. To better understand how we got to where we are today, we need to take a quick look over our shoulder at the way industry itself developed.

The Evolution of TQM

1750–1900: The Industrial Revolution

Until the Industrial Revolution began in the mid-eighteenth century, most items were custom-made. Industrialization brought about a fundamental shift from cottage-industry production to large-scale manufacturing. Simultaneously, industrial activity underwent extensive mechanization.

With the introduction of the assembly line between 1900 and 1940, products passed consecutively through various operations. Largely as a result of Henry Ford's influence, standardization became the order of the day. Ford's approach to pleasing his customers was famously captured in the phrase, "They can have any color they want, as long as it's black."

The prevailing management thinking of the day was based on the assumption that people dislike work, that money was the only motivator (except perhaps for fear), and that few people are capable of self-direction. Policies centered around keeping jobs simple and under close supervision. The expectation was that workers would meet standards only if closely controlled.

The World War II Years: America's Triumph

With the start of World War II, U.S. factories geared up for wartime production. Consumer goods grew scarce, and there was a loss of skills to the war effort. Factories were split into functional departments.

At the end of the war, the United States, stronger than ever after its war effort, undertook the rebuilding of Japan's war-shattered economy. Among the many Americans who were sent to Japan as part of that effort was Dr. W. Edwards Deming, a statistician. During his stay, Dr. Deming began to discuss the principles of industrial efficiency he had been working on. He achieved a major breakthrough in getting his message across when he addressed forty-five Japanese CEOs at a conference of the Japanese Union of Scientists and Engineers. They bought his message wholeheartedly.

Deming emphasized that business should know and respond to customer demands. Without customers, there would be no orders; without orders there would be no work. He advocated the use of statistical tools, such as surveys, to keep up with customer needs. He stressed the human aspects of management and the need to understand the way people work, think, and act twenty-four hours a day, seven days a week.

And finally, he advocated a climate of continuous improvement. "Listen to me," Deming told the Japanese, "and in five years you will be competing with the West. Keep listening, and soon the West will be demanding protection from you."

Deming did not tell the Japanese exactly how to do all this. They had to figure it out for themselves. And they did, using his general principles, which they applied first to manufacturing and then to sales and other areas. Slowly but surely, through continuous struggle and much trial and error, they began to make headway.

That they were not above slavish imitation along the way is illustrated by the following story. It was told to me by an associate who was working in the machine tool industry during this time and attended a lot of trade shows. In the midst of preparations for one particular show, his company's product display, a large machine tool, was accidentally dropped and the base cracked. Welders were brought in to fix the base so that his company could participate in the show. Several shows later, he noticed a Japanese version of his product. Each of the machine tools that were displayed by the Japanese had a weldment in the same spot that his display had. It was clear that the Japanese had learned to copy our product exactly.

1950–1970: U.S. Industrial Leadership

While Japan was still struggling to find its feet, the United States was emerging as the world's foremost industrial power. The U.S. controlled 75 percent of the world's manufacturing means, and by 1960 was producing 35 percent of the world's goods (compared to Japan's mere 2 percent). The U.S. controlled 95 percent of the global market in automobiles, steel, and consumer goods. This near-monopoly allowed U.S. manufacturers to ship product at 90-to-95 percent quality, a standard that was considered good enough. With U.S. products in high demand, guidance from experts like Dr. Deming wasn't considered necessary.

During those years, I personally remember being in staff meetings at which the quality people indicated that we needed to get to higher levels of quality. The general manager would ask the controller to determine the costs of achieving this, and when the costs turned out to be higher than the warranty reserve, it would generally be decided that such quality measures were not cost-effective. This was the way businesses were run at that time.

By comparison, consider today's environment, in which accepting even a 99 percent quality standard carries frightening risks. (Such a standard would write off as "normal," for example, some five thousand bungled medical operations each week, nearly two thousand too-short or too-long landings at major airports each day, and the dropping of seven thousand babies in maternity wards every twenty-four hours!)

Gradually, during the U.S. industrial heyday of the postwar years, a sea change began taking place in American management philosophy. Executives became converts to the idea that workers needed to feel important, needed to be recognized and consulted. As a result, managers began to discuss plans with their employees and listen to their objections and suggestions. This approach was expected to increase productivity as a result of *employee* satisfaction and participation.

If this sounds like an overly simplistic approach to productivity, don't forget that American life in general was simpler back in those days:

- Two-parent families were the norm, with Dad at work and Mom at home raising their 2.4 children.

- Americans were either married or unmarried; they didn't "cohabit" with "domestic partners."

- Americans either worked 9 to 5 or not at all; there was little or no part-time, flex-time, free-lance, or "consulting" work.

- There were fewer consumer options: phones were black, bathtubs were white, and checks were green; the hardest choices were between Fords and Chevies, chocolate and vanilla, *Life*, *Look*, and the *Post*, NBC, ABC, and CBS.

Meanwhile, Dr. Deming's ideas were continuing to influence the Japanese, who, while learning about quality, were also mastering the principles of repetitive manufacturing. As Deming himself noted, "Nobody except the Japanese seems to understand that, as you improve quality, you also improve productivity." By the 1970s, Japan was in a position to cherry-pick many U.S. markets. It had even begun to dominate some of them.

The 1970s: U.S. Industrial Decline

Meanwhile America's economic dominance had begun to decline. The U.S. share of the world manufacturing market was down to 17 percent, and U.S. productivity was experiencing decreases of up to 4 percent per year. Chrysler needed a taxpayer bailout to avoid bankruptcy; Ford's future was threatened by Japanese car makers. Xerox had experienced significant loss of market share to Japan; John Deere's reputation for quality was tarnished. The nation as a whole was losing its market dominance in cars, steel, motorcycles, radios, TV and video recorders, stereos, cameras, pianos, guitars, ships, and semiconductors—to name but a few of the most important areas.

American and European businessmen began making trips to Japan to learn the Japanese secret. The components of that secret, they learned, were the following:

- Lower wages
- "Quality circles"
- Automation
- Streamlining of repetitive tasks
- Morning exercises

Observers tended to lock onto one or another of these discoveries as a panacea for revitalizing U.S. industry. One prominent executive thought the esprit de corps generated by sharing morning exercises and singing the company song was the key to success. Many others became quality-circle enthusiasts.

While the Japanese hadn't even started using quality circles until 1965—fifteen years after they began to implement Deming's principles—the idea took hold among American CEOs that the technique had been central to Japanese industrial growth. Before long, individual divisions within companies were vying with each other to see who could create the largest number of quality circles—all in the desperate hope that this one technique would solve all their problems. As Dr. Deming commented regarding this phenomenon: "They didn't know what to look for. They didn't even know what questions to ask!"

There was nothing to see. The Japanese had created a management system that allowed them to build the right things right, the first time. They had figured out how to get everyone to work, think, talk, and act in such a way that things were done correctly all the time. That's how they had become the top auto producer in the world, making cars in 11 hours that U.S. manufacturers took 31 hours to make. That's how, between 1952 and 1970, they had lowered the cost of Japanese car production 41 percent, while the cost of U.S. car production had gone *up* 48 percent.

The 1980s: American Industrial Resurgence Begins

Not too long ago, the U.S. woke up to these dismal facts and figures and began a slow but steady climb back up. The resurgence of the 1980s, however, was undertaken in a vastly changed America:

- Forty percent of the work force was now made up of women.
- Many people were beginning to set up offices in their homes, relying on PCs and modems, fax machines and overnight carriers to keep them in touch.
- Products relating to health, fitness, and the environment were becoming big business.
- Deregulation and multiple-choice were the name of the game in every market from car models to cable stations, airlines to phone companies.
- The world of business had become a swarm of buzzwords and quick-fix solutions, with the argot multiplying faster than the junk bonds on Wall Street.

The field of purchasing was touched by all these developments—perhaps by none more so than by the "quick-fix," "miracle-of-the-month" phenomena. In purchasing, this was accompanied by a flurry of special terms and acronyms—some farfetched and forgettable, others right-on-target and of real value—as old paradigms were replaced by new ones and the stage was set for the evolution of TQM.

One of the first quick-fix solutions was Material Requirements Planning (MRP), which entailed forward-looking on parts requirements. This first, or "little," MRP was swiftly followed by a second, or "big," MRP, Manufacturing Resource Planning (MRP II), which entailed the linking of the little MRP with such other functions in the organization as receiving, purchasing, and accounts payable. Between them, the two MRPs were supposed to be the answer to all our problems.

Subsequently, "Just-In-Time" (JIT) purchasing, with its primary focus on suppliers, became the new panacea. But in too many cases, manufacturers were getting parts "just in time," at the expense of the suppliers, who had to carry all the inventory. A later version of JIT expanded the concept to any form of waste, time, material, capital, and so forth. Then came Computer Integrated Manufacturing (CIM), the automation and integration of processes.

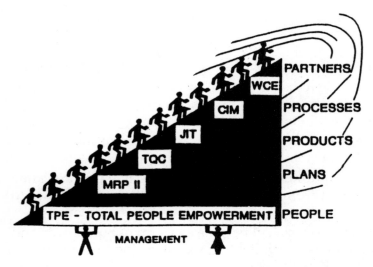

Figure 2-1. American resurgence—a nation of three-letter acronyms. (*Alban Associates, Inc.*)

Figure 2-1 shows how one company envisioned some of these three-letter acronyms as rungs on the ladder to success.

Despite the proliferation of quick-fix solutions, however, the right management structure was not yet in place for any method to be fully successful. The adoption of Total Employee Involvement (TEI)—the recognition of people and the importance of their involvement—provided the giant leap necessary to start the process of TQM—a process that eventually taught us how to take all these elements and develop the necessary philosophies and approaches to put them all together.

The management philosophy needed to implement TQM was based on the assumption that people *want* to contribute—that they themselves represent a major untapped resource. TQM policies were designed to create a climate in which all could contribute. The emphasis was on developing full participation on important issues, with the goal of bringing about improvements in decision making and control.

1986: The U.S. Defense Department's Assessment of Japanese Manufacturing Technology

In 1986 a special team called the Technology Assessment Team (TAT) was formed by the Department of Defense to study and evaluate Japanese manufacturing plants, technologies, and organizational skills. TAT made several visits to contractors in the United States as well as to contractors in Japan.

TAT found that manufacturers in the United States focused on achieving superiority in qualitative and quantitative product technology rather than on manufacturing practices and development. This focus was contributing significantly to the inroads being made by the Japanese in U.S. markets. TAT concluded that Japan's excellence in manufacturing was due to an underlying belief in the importance of manufacturing excellence, in the skillful exploitation of advanced technologies, and in wise management and organizational skills.

TAT identified six key elements in Japan's manufacturing performance:

1. Recognition of manufacturing as a strategic factor
2. Concurrent design of product and life-cycle processes
3. Quality
4. Continuous manufacturing-process improvements
5. Human resources
6. Suppliers

Recognition of Manufacturing as a Strategic Factor. The Japanese believe manufacturing excellence provides a key advantage in their national prosperity and continuous economic growth. In the early 1960s the government instituted the Ministry of International Trade and Industry (MITI). The main purpose of this program was to modernize the facilities of small manufacturers by providing five-year interest-free loans for 50 percent of the value of purchasing and leasing capital equipment. This continual stream of new, superior-quality, high-technology products is a major factor in the strategic planning of Japanese corporations. This investment occurs continually, not just when needed. Government has also increased engineering education so that manufacturing attracts the best and the brightest technical graduates.

Concurrent Design of Product and Life-Cycle Processes. This technique results in designs that have fewer parts and more standardization of those parts, enabling them to be more easily reproduced. The designs are relatively fixed, requiring limited engineering change orders, so that little line disruption results. This also enables new versions of popular products to be introduced with great speed and ease. Aggressive use of rapid prototyping by the Japanese yields early product and design stability, and is a key factor in controlling both direct and indirect costs. Accordingly, development time and manufacturing labor are also reduced. The learning curve is also accelerated, yielding fewer errors and higher quality.

Quality. Quality is achieved by the Japanese through a process of continuous improvement. They consider quality an integral part of product and process design. Quality is the concern and responsibility of every employee, from the president on down to the custodian. Top management is directly *involved* in total quality management—as contrasted with merely *supporting* it. Management directs and participates in quality education and establishes goals, while shop workers meet frequently with top management in quality circles on improvement projects. There are well over a million quality circles in Japan, each with an improvement theme the employees can relate to.

In the United States, 20 to 25 percent of production cost goes to the quality assurance personnel who find and correct mistakes. In Japan, only 3 percent of production cost is spent this way. The Japanese assign the in-process inspection to individual production workers who complete elementary statistical analyses and are authorized to take corrective action. This results in greater individual pride in workmanship and higher employee motivation. Only when unacceptable process variability cannot be solved by the worker is the problem referred to a higher level.

In addition to the efforts of individual workers, just-in-time delivery, flow-through manufacturing, and group technology help keep buffer stocks low so that motivation remains high. TQM also extends to a company's suppliers. This standard of high quality makes automation practical. Japanese quality has achieved error rates of one nonconforming part per million and one assembly error per quarter million.

Continuous Manufacturing-Process Improvement.   This element has to do with the almost painfully close attention the Japanese pay to detail.

M. Diamond, a president of ITT Defense, relates an illuminating experience he had during a visit to Japan. While touring a Japanese manufacturing plant, he heard the noise of an oncoming cart. The driver was really moving, and when he applied the brakes at his stop, there was a loud screech. The man jumped off the cart, ran to a machine, moved parts from one machine to the next, loaded the cart with the remaining parts, jumped on, and sped off burning rubber. As he was driving off, Diamond noticed that attached to the back of the cart was a broom that swept the floor—a small but powerful comment on Japanese efficiency.

The Japanese approach is to design a flexible facility and then constantly improve that facility by implementing small changes that increase productivity, efficiency, and quality. New state-of-the-art manufacturing methods are developed by in-house engineering laboratories and implemented by in-house automation departments. Production engineers often move to the factory and operate the line for transitional tests and training. Factory workers are involved in the process from the earliest stages.

Material flow has been simplified by producing only the amount of material needed for the next manufacturing step. This eliminates stockpiling, inventory carrying costs, and warehousing. The Japanese can smoothly scale volume up or down and rapidly change the product mix.

Human Resources.   People are Japan's most important asset. Manufacturing in Japan is fundamentally a human-centered process. Managers believe that their workers contribute both to manual and intellectual tasks. Blue-collar workers are graduates of senior or technical high schools often equivalent to our junior colleges. Japanese engineers are also well educated, and Japanese factories have about four times as many engineers as U.S. factories. The high level of education in the manufacturing facilities allows the Japanese to accelerate the pace of automation. Company-sponsored education and training is the norm rather than the exception. Individual education and training goals are established for each employee and their progress is reviewed periodically by senior supervisors. Japanese workers are not threatened by new technologies, nor do they require extensive retraining when new systems and processes are introduced. In fact, the workers may have aided in the design of the new system.

Workers share responsibility and authority with management. They are not chastised for stopping production, but rather commended for noting problems or imperfections. Job rotation is also practiced among both white- *and* blue-collar workers. Managers and engineers rotate about every three years, while floor workers are encouraged to rotate and learn all the various types of machines. As a direct result of the emphasis on education, compensation, and personal pride, the Japanese have impressive job stability.

Suppliers. The sixth and final element in Japan's manufacturing excellence is the network of suppliers that a business can deploy to take advantage of changing opportunities. TAT did not believe that Japanese suppliers were any more productive than their U.S. counterparts. But the evidence suggests that a major difference between U.S. and Japanese manufacturing may lie in the area of what is called *interfirm productivity:* the total productivity of a company and its suppliers.

The Japanese make widespread use of secondary subcontracting. Firms only deal directly with a few major suppliers. Relationships with subcontractors are long term and based on quality and performance in addition to price. Each supplier is highly specialized and very efficient, perhaps because there is an extensive exchange of technical data among subcontractors and the core firms.

TAT Recommendations

As a result of the study, TAT had nine recommendations for the U.S. Department of Defense:

1. Consider how policies that affect the industrial base can be co-ordinated with those of other government agencies, such as the Treasury and Commerce departments, to achieve maximum benefit.

2. Recognize the strategic importance of manufacturing and change acquisition policies to acknowledge the quality of the production process as a critical factor. Develop a better relationship with the industrial base.

3. Increase incentives to reduce the financial risk to industrial investments in new manufacturing equipment and techniques. (Particular attention should be paid to modernization of the supplier or vendor base.) Such incentives could include: budget and contracting policies that encourage stability for the contractor and the suppliers; rewards to contractors that continually use state-of-the-art production systems; changes in accounting systems to capture the benefits of production modernization.

4. Actively focus on developing policies that will encourage the use of concurrent designs. Develop a process that reduces reliance on manufacturing specifications and encourages design and process innovation.

5. Create policies that encourage quality to be *designed* in, not *inspected* in.

6. Establish a research budget to anticipate unique design, manufacturing, and field service problems for future requirements.

7. Take an active part in ensuring the presence of qualified and educated personnel in the industrial base.

8. Establish policies that encourage prime contractors to establish long-term partnerships with their supplier base. Devote more effort to getting suppliers involved in the design of the product.

9. Require TAT to continue its study of the Japanese manufacturing approach. Form additional study groups to compare Japanese manufacturing with that of other countries.

1987–1988: MIT Study Identifies Weaknesses in U.S. Industry

Americans have conducted many surveys and studies to understand what U.S. industry needs to do to be successful. In 1987–1988 a team of sixteen senior professors at MIT (the MIT Commission on Industrial Productivity) conducted research into eight major industries on three continents to identify the key factors that have an impact on quality and productivity. The study identified six recurring patterns of weakness in U.S. industry:

1. *Outdated strategies.*   Our reliance on the mass production of standard commodity goods has blinded us to the growing strength of scientific and technological innovation abroad.

2. *Short-term profit orientation.*   U.S. manufacturing needs to put less stress on short-term profits and more emphasis on quality, productivity, and product development.

3. *Technological weaknesses in development and production.* U.S. firms find it difficult to design simple, reliable, and manufacturable products. We fail to pay enough attention at the design stage to reduce problems in manufacturing.

4. *Neglect of human resources.*   Major reforms are needed in the American educational system. On entering the business world, our young people are far behind those of other societies in mathematics, science, and language attainment. In the workplace, training is limited and amounts to little more than "following Joe around."

5. *Failures of cooperation.*   There is a lack of cooperation at several levels: between individuals and groups within a firm; between firms and their suppliers and customers; among firms in the same industry; between labor and management; and between firms and governments.

6. *Government and industry at cross purposes.*   U.S. industry has experienced problems not so much with the *amount* of government intervention as with the *type* of government intervention. Regulatory strictness has caused the most trouble.

The commission's report (published under the title *Made in*

America: Regaining the Productive Edge, by MIT Press, Boston, 1989), identified five imperatives for a more productive America:

1. Focus on producing well: put production ahead of finance.
2. Cultivate a new economic citizenship: create an involved, educated, responsible, and rewarded work force.
3. Promote the most productive blend of cooperation and individualism.
4. Learn to live in the world economy.
5. Provide for the future: invest in education and save for productive investment.

The 1990s: Is the Japanese Industrial Threat Receding?

The American Society for Quality Control has recently issued the results of a 1989 survey indicating that Japan is no longer perceived as the biggest challenge to American business. The survey polled 601 companies from the Fortune top 1000 companies as well as numerous smaller companies. Their findings were as follows:

- 62 percent of these companies stated that their greatest competitive quality challenge now comes from other U.S. companies.
- 9 percent indicated that Japan was still their greatest challenge.
- 51 percent said that the U.S. is gaining in quality on foreign competition.

If all this is correct, we may have come full circle.

Can We Be Successful at TQM?

One school of thought holds that we in the U.S. cannot be successful at implementing TQM. Among the reasons cited for this view are the following:

- *TQM isn't sexy.* Concepts and strategies like Manufacturing Resource Planning (MRP II), Computer Integrated Manufacturing (CIM), Statistical Process Control (SPC), and, especially, Just-In-Time (JIT) purchasing are more attractive and appealing as strategies than Total Quality Management. TQM just doesn't cut it and will not demand the same kind of attention.

- *We are too impatient.* We aren't willing to wait the necessary five to ten years to see the real results of our efforts. Our senior management will focus only on those activities that will ensure year-end bonuses. Hardly anyone is prepared to embark on a policy requiring all new employees to get two years of training before their first permanent assignment.

- *TQM offers no magic solution.* We like cookbook solutions to problems, and TQM is a new and evolving process. There aren't too many TQM solutions that are a step-by-step proven process. What works for one company may prove unsuccessful for another. There are no set rules to follow; each attempt to implement TQM is an individualized process. Guidelines are available but are just that, guidelines. The implementation of TQM requires trial and error, and initial attempts can prove discouragingly unsuccessful.

- *Our role model is Superman.* We often approach problems the Superman way—by throwing everything we have at them to put the fire out fast. The Japanese approach problems more like Detective Colombo. He pays attention to the details, digging and digging until the real culprit is found. The Japanese dig and dig until the root cause of a problem is found, and then solve the problem. This prevents recurrence.

- *You can't change a company's culture.* Our ways of doing things are too deeply ingrained for us to be able to change, especially when it comes to the way we conduct business. Our employees won't make the necessary changes.

Another school of thought holds that U.S. industries *will* be successful at TQM, and have, in fact, already had some success.

- *We are not starting with a blank page.* We have already done and are continuing to do a number of good things. All we need to do is build on those successes. After all, aren't we the ones who taught the Japanese?

- *We can be terrific team players.* TQM requires teams for implementation. The United States has a lot of experience with teams: just look at all the sports we play. The U.S. has more teams than any other nation.

- *Government involvement.* Both federal and state governments are involved in the TQM effort. The Malcolm Baldrige National Quality Award from the Department of Commerce recognizes the highest-quality companies in the country. October has been designated National Quality Month. State governments are sponsoring similar kinds of quality-oriented activities and awards.

Whichever view one holds, the current trend is to implement programs that will generate better quality than we have achieved before. Purchasing programs will be a major player in that activity.

Six U.S. Companies That Are Already Successful at TQM

Some U.S. companies are already responding to the weaknesses cited in the TAT, MIT, and ASQC studies. A sampling of such companies includes the following:

A. T. Cross

AT&T Network Operations Group

ITT Defense

Motorola

Texas Instruments

Westinghouse's Commercial Nuclear Fuel Division

A. T. Cross gives every production employee the power to reject any imperfect part at any time during the production process. After the product leaves, it is protected by the company's lifetime mechanical guarantee.

AT&T Network Operations Group had each of its officers and directors review three "resolved customer's complaints" and meet with the customer to see if the customer was truly satisfied with the resolution. The executives were not to reveal that they were officers and directors. The results were eye-opening. Many customers were not at all satisfied with the solutions. They complained about broken systems and processes and a lack of focus on the customer by AT&T employees. These findings provided the basis for a new quality improvement process.

ITT Defense's TQM effort focuses on meeting the needs of the customer. ITT believes that this will best allow them to continue to be successful in their mission. The ultimate goal of ITT's TQM is that it will cease to have its own identity, that the label "TQM" will eventually fade from use as continuous improvement philosophies and principles become the normal business approach.

D. Travis Engen, former president of ITT Defense, spearheaded ITT's commitment to TQM. In 1988, he initiated a survey of quality consciousness, beginning with a questionnaire to nearly all locations and all levels of the company. The questionnaire was followed by group discussions and hour-long individual interviews of several dozen employees by an outside consultant.

The results of the quality survey were reviewed by the president, all unit general managers, and the ITT Defense staff. The responses were positive. ITT Defense employees believe that their supervisors are interested in their suggestions for improvement. They also say of themselves that they are an organization with a very strong response to the customer. The session generated the following vision of ITT Defense:

WE ARE ITT DEFENSE

- We are the best in the world at what we do.
- We take pride in our technology, teamwork, and integrity.
- Our employee, customer, and supplier teams strive for continuous improvement through mutual trust and open communication.
- Our customers rely on us to provide the best, and we deliver.

The session also established the management and cultural environment that would encourage and accept changes. Management at ITT accepted the necessary up-front investment and long-term commitment to achieve a TQM philosophy. It committed to creating a new and more flexible environment and culture that would encourage and accept change.

In addition, an executive level TQM steering committee was formed. The committee, comprised of representation from ITT Defense and unit general managers, has responsibility for developing and maintaining TQM philosophy throughout the company.

Dr. Genichi Taguchi, world-renowned engineering expert and management consultant for optimizing the design of processes and products, credits ITT with becoming the "leader in Taguchi applications in the Western world." ITT sponsors an annual symposium for industry and government.

Elements of TQM principles and practices have been in use at ITT since 1985. ITT recognized that a systematic approach to continuous improvement was essential in order to meet increased production deliveries resulting from (1) a projected fourfold increase in business base, and (2) a shift from engineering to manufacturing. To that end one of the divisions, Avionics, initiated an operations strategy: a process that developed a detailed analysis of present operations ("as-is analysis"), established an optimized business model as a goal for the future, and developed and implemented plans to lead AV toward this goal. In the process, key opportunity areas were identified and cross-functional action teams were established. Since its inception, over seventy-five individual projects have been completed. The major TQM emphasis has been on the application of quantitative methods such as the Taguchi Method, Statistical Process Control (SPC) and Engineering to Production transition.

ITT adopted the DoD model 5000.51g (see Figure 5-1) as a guide for implementation. The model is primarily an improvement/project/creation model. It has seven steps that begin with establishing a TQM cultural environment and result in implementing a continuous cycle of improvement projects aimed at improving organizational performance.

ITT uses cross-functional teams called Critical Process Teams (CPTs). CPTs are used to focus on top-level business ideas, such as company system requirements, or management-identified problems. Team members are currently selected by the committee and consist of members from appropriate departments. CPTs are organized to last only as long as necessary for successful implementation of an idea or an improvement. The use of CPTs not only brings about process improvement, but also breaks down traditional organizational barriers and fosters ongoing, two-way communication among employees in various disciplines.

ITT's approach to training includes a cascading effort. Each level of management is required to train the next level. This assures that the principles and techniques are learned by every employee.

Motorola's naming as one of the recipients of the first Malcolm Baldrige Award was a recognition of its commitment to quality. Its formal program of Six Sigma Quality (a target of no more than 3.4 defects per million products) has become a model for many companies. An example of Motorola's efforts is its focus on reducing total cycle time. To meet competition in developing a new pager in 1985, Motorola realized that it would have to bring the product to market in eighteen months, compared to the standard three-to-five years. To accomplish this, the engineers had to modify existing products (rather than starting from scratch), while the manufacturing group studied their competitors' processes. The result of their efforts is that today, two hours after an order for a custom pager is entered into computers in Schaumburg, Illinois, the pager comes off the line in Boynton Beach, Florida.

Texas Instruments has invested heavily in technical improvement techniques. One of these is Statistical Process Control (SPC), a methodology for measuring and controlling variances in operating processes. Another is Quality Function Deployment (QFD), a technique for identifying customer needs (see Chapter 7 for the use of QFD in purchasing). One story, told by Bob Porter, vice president for quality and reliability for the Materials and Control Group headquartered in Attleboro, Massachusetts, comes from Texas Instruments in Italy. Engineers there had come up with a new product that they thought would allow them to leap-frog their competitors. By the time they completed the QFD analysis, it was concluded that this design had only a marginal advantage. As a result, the product was scrapped prior to the purchase of machines and dies and the hiring of staff.

Westinghouse's Commercial Nuclear Fuel Division (CNFD) was rewarded for its performance record in the teeth of the debate about the nuclear industry. For delivering product that was 99.995 percent flawless, CNFD was a first-time recipient of the Malcolm Baldrige Award. CNFD uses Westinghouse's highly regarded Productivity and Quality Center, established in 1979. The center was the first corporate-sponsored resource to be devoted to quality improvement, with 130 dedicated quality planners, engineers, and consultants on its staff.

In 1990, Chairman of the Board John Marous instructed all ninety corporate divisions to compete for the George Westinghouse Total Quality Awards, two internal prizes modeled after the Baldrige Award. The best unit and the most improved unit each receive $200,000 to spend on anything "as long as it is not immoral or illegal," says Marous.

Listing the six lessons he has learned about this vital subject, Marous states that quality:

1. Is a matter of survival
2. Requires a cultural change
3. Takes time
4. Demands that top management pay attention
5. Needs a scoreboard
6. Must involve everyone

How Do You Know If You've Made It?

The ability to measure the success of TQM varies. With all the bankruptcies being experienced nationwide, it's easy to believe that if you wake up and find yourself still in business, you are successful. Other measures of success include:

- Winning the Malcolm Baldrige Award, or at least a site visit. Those few companies that have received this prestigious award are considered successful by many.
- Believe that quality = profits. When your company truly believes that improved quality does result in improved profits, you have passed the first milestone on the road to success.

- Honeywell claims success because the employees' annual performance review includes team input.
- Motorola's success ties into their policy that requires the chairman of the board's approval to dismiss an employee with ten years of service.
- The president of General Dynamics indicates that he will have success when he stops receiving anonymous letters.
- A company can claim success when and if it becomes the supplier in the following story: A CEO of a medium-size manufacturing company attended several executive seminars and decided that his company would implement total quality management. He assembled his staff and made the announcement. He also stated that because he had learned in the seminars that TQM is driven from top down, he would give specific instructions to each of his vice presidents. When the purchasing executive's turn came, the CEO gave these instructions: "Starting immediately, all purchase orders we send to suppliers will carry a statement to the suppliers indicating that we want only 1 percent defects." The vice president of purchasing and the vice president of quality tried to argue against this, but the CEO prevailed. Several weeks later a package from one of the suppliers arrived with a note attached to a bag of components. The note was addressed to the company and stated that enclosed were the 1 percent defects you asked for. "We don't know why you wanted them, but here they are," the note said. On the back of the note, the supplier had also written, "and by the way, they were hard to make."

3

What Does TQM Mean to Purchasing?

The bottom line for purchasing is that, increasingly, it is going to be required to deliver higher quality, more reliable product, and more often—with the same or fewer resources—as it does today.

Introduction

Purchasing is a key ingredient in the implementation of TQM for a very simple reason. To see why, take a look at Figure 3-1. This study by General Electric shows that the cost associated with finding a problem during inspection of a product at delivery is just 3 cents, while the cost of finding the same problem after the product has been installed is $300.

In the 1970s it was considered good business to sell a product requiring recurring service. (Remember the term *planned obsolescence?*) In today's climate, however, such a product will not sell— or will not sell more than once.

Senior executives consider the role of purchasing critical to improving quality. In a survey conducted in 1989 by the American

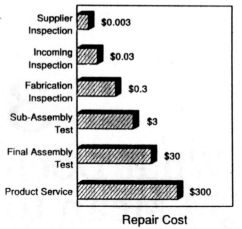

Figure 3-1. Fault-repair costs. (*General Electric*)

Society for Quality Control, 601 senior executives indicated that more control over suppliers was one of the top ten areas in which quality could be improved throughout American business. (Other target areas were employee motivation, employee education, process control, and quality improvement teams.)

While most companies understand that 60 to 70 percent of the cost of most manufactured products is materials, we have only recently realized that the cost of buying materials and services is more than just the unit price. Figure 3-2 illustrates the many components of the total cost. The other components can amount to as much as, if not more than, the unit price. Moreover, the concept that the market dictates the price and the buyer has little influence may be appropriate for the unit price, but buyers *can* influence the other parts of the total cost. This is another reason why there is a new focus on purchasing in TQM.

Suppliers and customers alike have never been as motivated as they are today to improve the quality of products and services to a level previously believed impossible. In many companies the burden of achieving this new level of supplier quality has been placed directly on the shoulders of the purchasing department.

The experts tell us that the solution is for purchasing to become more efficient and to have greater supplier involvement. This

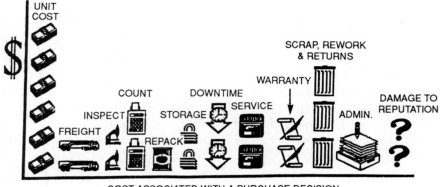

UNIT
COST

SCRAP, REWORK
& RETURNS

WARRANTY

COUNT   DOWNTIME
SERVICE   DAMAGE TO
INSPECT   STORAGE   REPUTATION
FREIGHT
ADMIN.
REPACK

COST ASSOCIATED WITH A PURCHASE DECISION

Figure 3-2. The "color of money"—costs associated with a purchasing decision.
(*Coopers & Lybrand*)

kind of general guidance is great, but how should it be applied?
To understand this, it is appropriate to look at how purchasing is
changing under TQM.

Several traditional elements of purchasing are changing:

- The role of purchasing
- Acquisition strategies
- Buyer/supplier relationships
- Purchasing performance measurements

The Role of Purchasing

TQM is changing purchasing's traditional role. Before TQM, the
supplier viewed the buyer as the customer. In this approach there
existed a balance of power. The supplier usually knew more
about the product it sold, while the buyer made the decision in
selecting the supplier. Under TQM, the buyer views the supplier
as the customer. The buyer is required to provide the supplier information, such as drawings, specifications, and required dates,
that will allow the supplier to provide products or services as
needed. This new role gives the supplier an unfair advantage, but

the risk to the buyer is offset if the supplier and the buyer share mutual goals and benefits.

Before the advent of TQM, the buyer's role was to select suppliers and place orders. Under a TQM system, the buyer's role is to serve as the team leader of multiple functions within his company (engineering, quality, and manufacturing). The team is charged with supplier selection as well as with providing the supplier with appropriate training. Training will help ensure that the goods and services provided meet the needs of the buyer.

Acquisition Strategies

Acquisition strategies have also changed:

Traditional Acquisition Strategies	TQM Acquisition Strategies
Use multiple sources	Qualify supplier base (reduce/expand)
Market dictates price; buyer has little influence	Buyer looks at total cost, not just at price
All technologies developed in-house	Buyer uses suppliers' technology
Suppliers get drawings and determine whether they can supply the parts or services as specified or must request changes in specifications	Suppliers are involved in design

The illusion of multiple sources leading to assurance of supply has been replaced by reduced-source or single-source strategies. The trend today is to reduce the supplier base and to use multiple sources only when a unique technology is required or when one source cannot provide all the parts. The strategy of Ford Motor Company, for example, is to rely on a single source for parts wherever possible. A notable exception in Ford's case is tires, since one tire company cannot supply all Ford's needs.

The buyer's influence on unit price is limited. The market tends to determine price levels, so that transaction is little affected by the buyer-supplier relationship. However, other elements—total cost, freight, downtime, and so forth—can be heavily influenced by the buyer's relationship with the supplier.

The concept of vertical integration (developing all technologies in-house) is no longer economical for most companies. The complexity of many of our products today requires more use of our supplier's technology.

The use of suppliers in the design stages of a product not only takes advantage of the suppliers' technology but also provides an opportunity for supplier input. This input allows a supplier to effectively utilize the supplier's manufacturing processes in the final design of the product.

Supplier Relationships

TQM's greatest impact on purchasing has been in the area of supplier relationships.

Traditional Buyer/Supplier Relationships	Buyer/Supplier Relationships in a TQM Environment
Buyers and suppliers work in constant conflict (a win-lose situation)	Buyers and suppliers work in partnership (a win-win situation)
Buyers are always considered right (implying that suppliers are always wrong)	Suppliers are considered experts in the commodity they supply
Buyer-oriented supplier asks: "What can I do to get the business?"	Supplier-oriented buyer asks: "What can I do to help you become a better supplier?"

Developing partnerships does not eliminate the need for good negotiations. However, the objective of negotiations is for both parties to benefit. The old tendency, when problems with supplier's products arose, was to blame the supplier. Today, the approach is to find out how each party can help the other to improve performance. The buyer's ongoing role is to help ensure that the supplier has the right tools to provide good services and products.

Purchasing Performance Measurements

Purchasing has a second customer, the requisitioner. Purchasing has always been considered an internal service organization, but within a TQM environment the emphasis on meeting all the customer's needs is more pronounced.

Under the TQM umbrella, assessing purchasing's performance includes not only measuring the efficiency of the individual buyer and of the purchasing department (and determining the cost of issuing a purchase order), but also measuring how well the needs of both the internal customer (the requisitioner) and the external customer (the supplier) are being met. The latter assessment involves measuring the control of the supplier base, a process that develops supplier's resources to achieve a competitive advantage in the seller's marketplace. The resources include the supplier's technology, processes, quality, and cost. The following table compares purchasing performance measurements within a traditional environment and within a TQM environment.

Traditional Measurements of Purchasing's Performance in Meeting the Needs of the Requisitioner (the Internal Customer)	TQM Measurements of Purchasing's Performance in Meeting the Needs of the Requisitioner (the Internal Customer)
Requisition cycle time	Requisition cycle time
Meeting required dates	Meeting required dates
Quality of product/service	Quality of product/service
	Responsiveness to requisitioner

The priorities of the requisitioner under TQM include not only receiving products and services of sufficient quality on time, but also being able to count on purchasing for a response to specific needs. For purchasing, this means being available to answer questions in a timely manner and giving the requisitioner access to such supplier resources as technology and product/service capabilities.

Measuring the control of the supplier base includes meeting the needs of the supplier in addition to measuring the supplier's performance regarding quality and delivery.

Traditional Measurements of Purchasing's Performance in Meeting the Needs of the Supplier (the External Customer)	TQM Measurements of Purchasing's Performance in Meeting the Needs of the Supplier (the External Customer)
Sufficient data	Sufficient data
	Training of supplier

In addition to providing information to enable the supplier to provide the product/services needed, purchasing must also give the supplier training in process improvement and continuous quality improvement.

In a TQM environment, measurement of the supplier's performance covers additional areas.

Traditional Measurements of a Supplier's Performance	TQM Measurements of a Supplier's Performance
On-time deliveries	On-time deliveries
Quality of product/service	Quality of product/service
	Number of problems solved
	Responsiveness to inquiries
	Control of processes

In a TQM environment, the assessment of a supplier's ability to provide quality products in a timely manner is supplemented with an assessment of the supplier's performance in solving customer problems (such as technology needs). Also measured is the ability of the supplier's manufacturing processes to provide products with reduced variation in specification limits—i.e., products that are consistently made to conform to a specific target, such as .5 mil., etc.

The best supplier I ever had was one that not only delivered quality products on time but also tried to keep me out of trouble. He advised me in advance when he thought I should reorder, based on his records and on changes in technology that he thought might have an impact on the products he provided my company.

4

What Are the Leaders Doing?

Introduction

Purchasing professionals nationwide are aggressively pursuing programs to improve supplier relationships. However, some of them suggest that the whole idea of building such networks is in danger of becoming a meaningless slogan. Nevertheless, new ideas can be generated and lessons learned by reviewing the approach taken by several companies. These companies have had success with various actions resulting from the changing roles of purchasing.

Although each company's approach to TQM and suppliers is different, there are common themes that all seem to adhere to:

- Benchmarking provides the goals to strive for.
- Reducing the supplier base and heading toward single sources of supply provides a better vehicle for improved quality.
- Trust is the one element that will cement the relationships established.
- Investment in the training and technical support of suppliers is key to success.

- Focus on continuous improvement must become a way of life.
- Involving suppliers in the product design process yields the greatest benefits.
- Using metrics to measure performance is necessary to determine whether improvement is occurring.
- The sharing of information, risks, and rewards with suppliers is the only way to establish long-term mutually beneficial partnerships.
- Multidiscipline involvement in supplier selection and relationships strengthens the relationship.

The following companies will be discussed below in alphabetical order:

Apple Computer	ITT Defense
Bell & Howell	Kurt Manufacturing
Black & Decker	McDonnell Aircraft
Briggs & Stratton	Morton & Co.
Caterpillar	Motorola
Chrysler	Navistar International
Corning	NCR
Cummins Engine	Outboard Marine
Delta Faucet	Raytheon Small Missile Division
Ford	Thiokol
Harley-Davidson	Toyota
Haworth	Westinghouse
Honda	Xerox
IBM-Rochester	Yale Materials Handling

Apple Computer

Apple Computer's purchasing people firmly believe their success can be traced to their relationship with their suppliers. This claim is supported by contributors to the *Harvard Business Review.* Apple focuses on purchasing's involvement in the design cycle. This ensures that the best possible integrated circuit can be devised to provide leading-edge technology for high-quality product with short life cycles.

The tightness of Apple's supplier relationships can sometimes be measured in hours. Its purchasing staff claim their suppliers receive hourly updates on Apple's needs to facilitate delivery. Apple has five objectives for purchasing:

1. *Quality.*   With emphasis on design, supplier, and buyer's process.
2. *Velocity.*   Defined as the competitive use of time and rate of change.
3. *Cost improvement.*   Total life-cycle cost throughout design, development, test, and manufacturing process.
4. *Technology.*   Purchasing has a lot of input into design.
5. *Risk management.*   People are taught to calculate risk.

Bell & Howell

Bell & Howell focuses on certification to reduce its supplier base. Suppliers rated well in a preliminary assessment are invited to submit a detailed, written quality plan. These plans are analyzed to determine if they complement Bell & Howell's own business plans. An on-site audit is then conducted.

Certification means a multiyear partnership of up to five years, and in some cases certified suppliers will handle 100 percent of the business in certain commodities.

Black & Decker

Black & Decker has a rating system that is the primary basis for supplier selection. The system features quantitative ratings in four primary areas: quality, technology, delivery, and commercial. The program includes supplier on-site evaluations by Black & Decker in-house experts in the four areas.

Briggs & Stratton

A supplier quality engineer is assigned, sometimes full-time, to a supplier that does not conform to Briggs & Stratton's *Supplier Quality Manual* and whose manufacturing systems are not capable of producing the caliber of product required. With ten to twelve suppliers participating, the results have been the resolution of problems in one-quarter to one-half the normal time.

Briggs & Stratton has also created a new position, manager of supplier development, and launched a Certified Supplier Award based on quality performance, pricing, and technical capability.

Caterpillar

Caterpillar has a Quality Institute that has educated one thousand individuals representing four hundred suppliers.

Chrysler

The objective of Chrysler's program is to achieve an overall performance level of zero defects. Included in its program are the following elements:

1. All buyers attend in-house training four hours a day for three months.
2. Purchasing achievement teams nominate individuals who have excelled at their jobs and have contributed to Chrysler's goals.
3. Awards, ranging from plaques of recognition to month-long assignments of executive parking spaces, are given to reward performance.
4. Long-term contracts are established with suppliers. Twenty-

seven suppliers are asked to quote on two-to-four-year requirements.

5. The supplier base has been reduced by 20 percent.
6. Key suppliers are informed of short- and long-term plans and strategies.
7. All suppliers are advised that their prices may not increase.
8. Suppliers are asked to indicate how Chrysler requirements have cost them money and to suggest ways to reduce costs. Many suggestions have been the same as those coming from internal engineers.

Corning

Corning relies heavily on the Malcolm Baldrige criteria for their purchase program. The Baldrige Award establishes benchmarks and provides guidelines for companies attempting to develop their own internal programs and supplier rating systems.

Cummins Engine

Cummins sets targets for suppliers to shoot for and then helps suppliers reach these goals. The targets are established in four key areas: quality, technology, piece price cost, and administration.

The administration category includes the supplier's account services, frequency of visits to customer plants, and electronic communications capabilities that allow the supplier total access to Cummins' computerized scheduling bulletins.

Once a year each supplier makes a formal presentation to Cummins, assessing its progress toward the specific objectives set for that company.

Delta Faucet

In its first year, Delta Faucet's supplier performance program has developed partnerships with 80 percent of their suppliers that do at least a $40,000-a-year business with Delta. The program assesses quality, acceptance, delivery, and price. Under the partnership approach Delta forecasts up to a year's orders with some suppliers, with the first two months firm.

Ford

Ford spends $50 billion worldwide per year and has an impressive performance record. Thus far, Ford has

1. Globalized its supply base by establishing worldwide centers of excellence in design, engineering, and purchasing. To respond to global requirements, some suppliers are forming joint ventures or business associations with former competitors

2. Optimized its supplier base by trimming the base by 40 percent

3. Placed 70 percent of its North American suppliers under long-term contracts

4. Placed practically every production part for a given model single-sourced under a long-term contract

5. Focused on early supplier involvement and awarded 70 percent of its purchases to suppliers that assume concurrent engineering responsibilities for components

6. Taken a highly centralized purchasing approach, with the body and assembly plants doing very little of their own purchases except for some blanket orders and maintenance

7. Established the American Supplier Institute to bring suppliers up to date on new quality techniques, notably those based on the ideas of Japanese quality guru Genechi Taguchi

Since 1980 Ford's quality has increased by nearly 70 percent and reportedly leads the Big Three with the fewest defects per car. Ford was awarded the 1990 Medal of Professional Excellence by *Purchasing* magazine.

Harley-Davidson

Harley-Davidson's program focuses on generating savings. It has

1. Developed and conducted seminars for suppliers
2. Decentralized its purchasing department and located buyers in the shop areas close to the product and the people
3. Increased the number of buyers from five to nine
4. Requested from suppliers a 2-percent price reduction and an extension of terms to 60 days
5. Trimmed its supplier list in half—from 820 to 415
6. Arrived at single-sourcing for each part
7. Involved suppliers in product engineering
8. Allowed suppliers a 16-week window to the production schedule

The results have been an increase in inventory turns from $4\frac{1}{2}$ to 23 and an increase in market share from 23 percent to 40 percent.

Haworth

Haworth has developed unique relationships with its suppliers. The supplier designs and manufactures a product and Haworth markets it. This relationship is cost-effective for Haworth, which can develop more products with fewer Haworth people. The relationship is a source of more business for the supplier and allows the supplier to concentrate on manufacturing.

Haworth uses capability ratios and statistical formulas (Cpk) that hone in on the first piece and the capability of the supplier's tooling.

Honda

Honda has the top-selling car in the U.S., the Honda Accord. Honda's program:

1. Requires purchases to be 30 to 100 percent of suppliers' volume

2. Dedicates 120 engineers to deal with incoming parts and supplier-quality issues
3. Dedicates 40 engineers to work with suppliers to improve productivity
4. Helps suppliers develop employee involvement programs
5. Dedicates technical support to suppliers
6. Assists suppliers with business problems (i.e., rapid growth)
7. Conducts a loaned executive and a guest engineer program
8. Conducts frequent supplier visits

As a result, 40 current suppliers ship 100 percent defect-free, 100 percent of the time, and virtually all parts go directly to the assembly line. There is no inspection of the product when it is received at the buyer's dock or receiving department.

IBM Rochester

IBM replaced competitive bidding with competitive supplier evaluation. Suppliers are evaluated on:

- *Quality posture.* An evaluation of the supplier's quality system and SPC capability
- *Technical expertise.* A match of technical expertise with IBM's needs
- *Delivery.* A capacity for continuous flow basis and a willingness to reduce inventory and minimize product lead time

IBM insists that its suppliers use SPC and look for suppliers that have adopted Malcolm Baldrige principles. Although IBM doesn't require its suppliers to apply for the award as Motorola does, IBM has asked its suppliers to work with IBM in doing self-assessments.

IBM has already cut its supplier base by 35 percent in the last five years and plans to continue to reduce the base. IBM has moved toward suppliers delivering parts directly to the line and has tried to reduce lead times by having suppliers close to home.

In 1990 buyers received 40 hours of training. It is expected that in 1991 each buyer will receive 80 hours of training. IBM has developed a Certified Purchasing Manager (CPM) training program to help buyers get CPM-certified.

The new purchasing approach is credited with helping IBM win the 1990 Malcolm Baldrige Award.

ITT Defense

ITT utilized Quality Functional Development (QFD) to determine its customers' needs and select appropriate purchasing strategies (see Chapter 7). In addition, ITT developed a process to assure that suppliers were ready to meet its needs. The process is known as Supplier Readiness Assessment and is described in Appendix F.

Kurt Manufacturing

Kurt, a precision machine parts manufacturer in Minneapolis, Minnesota, focuses on getting information into the hands of suppliers as a key element in establishing good partnerships. The most important information falls into three categories: realistic quantities, accurate production start and stop dates, and correct market information.

McDonnell Aircraft

McDonnell Aircraft has a plan to reduce its supplier base from 2875 to 1000 with 500 suppliers designated as part of its Certification of Preferred Suppliers.

Certification consists of three distinct parts:

1. *Assessment of business processes.*   The elements assessed include management, quality, delivery, cost, technology, and customer support.
2. *Application of process control/SPC*
3. *Product performance.*   Supplier Performance Evaluation and Rating System (SPEARS) measures quality (the percentage free of quality defects), delivery (performance to purchase or-

der schedule), responsiveness (multidiscipline assessment), and price trends (at a part-number level).

A performance progress report is issued quarterly to suppliers. The report reviews the past 12 months' performance in terms of delivery and quality (including attention paid, during the final quarter, to fixing quality problems). The report also identifies items needing improvement.

Morton & Co.

Morton & Co., based in Wilmington, Mass., is a small company that manufactures precision parts. Its program has several key characteristics:

1. Open and consistent communications with suppliers
2. A comprehensive qualification program for evaluating and selecting suppliers
3. Long-term commitment to suppliers
4. An understanding of the total-cost concept
5. A can-do attitude

Motorola

Motorola is also a winner of the Malcolm Baldrige Award. Its purchasing program:

1. Benchmarks the Japanese
2. Has established a tenfold improvement in ten years
3. Has established a target of 6 sigma = 3.4 PPM defect rate
4. Requires all suppliers to develop plans for the Baldrige Award

Navistar International

Navistar has component management teams that employ a decision-analysis system to rate potential suppliers for many components of its trucks and diesel engines. The team consists of specialists from engineering, manufacturing, quality, transportation, marketing, and materials management (purchasing). The team completes mandatory training in a problem-solving and decision-analysis approach developed by Kepner-Tregoe Inc. of Princeton, New Jersey.

NCR

NCR's purchasing job is supply-line management. Purchasing is a key element of NCR's corporate overhaul, holding the reins on overall material expense rather than just on price, keeping the company technologically competitive, and also building relationships with suppliers of key technologies. The key qualifiers for NCR suppliers are quality and leading-edge technology. The following table illustrates NCR purchasing policies before and after the overhaul.

NCR Purchasing Policies Then	NCR Purchasing Policies Now
Contracts renegotiated yearly	Evergreen contracts
Incoming inspection	Benchmarking
Purchasing components	Securing technology
Over 2000 suppliers	150 suppliers
Spotty parts quality	50 to 100 ppm rates on incoming parts

SOURCE: *Electronics Purchasing.*

Outboard Marine

Outboard Marine's director of corporate purchasing, C. W. Cross, oversees buying for 11 domestic and four overseas plants and does central buying from 500 active suppliers. Outboard Marine's focus is on the development of an objective, quantifiable rating system that lets suppliers know where they stand.

Raytheon Small Missile Division

The manufacturer of the Patriot missile focuses on redoing its supplier base and training supervisors in techniques like statistical process control.

The 200-person purchasing department spends $300 million annually for 200,000 different production parts from 7000 suppliers. Its goal is to cut the supplier base by up to 40 percent by 1993.

Buyers measure the performance of suppliers and look for an acceptable parts-per-million defect rate, high performance in meeting delivery commitments, and quick response to problems.

The corporate group, headed by Marty Kane, director of corporate purchasing, evaluates the purchasing departments annually on the quality of incoming parts, delivery, the overall performance of suppliers, ethics, and the management of subcontractors.

Thiokol

Thiokol's supplier-partnership approach is a three-phased program:

1. *Long-term supplier relationships.* Thiokol applied this phase to Maintenance Repair Orders (MRO). The objective was to eliminate or reduce the cost of inventory, minimize the labor required to handle MRO suppliers, simplify invoicing, and improve customer service. The approach included getting Thiokol's internal users involved in the development of the program, briefing the suppliers on the program expectations, and issuing contracts for three to five years to successful bidders. The contracts are a three-year stockless supply contract with stationary suppliers, with items stocked at the suppliers. Suppliers deliver directly to each user within 24 hours. The user releases all shipments with no formal purchase order issued. Each user reconciles monthly deliveries and authorizes payment.

2. *New program partnerships using concurrent engineering to focus on manufacturability.* Competition is utilized on the front end to satisfy government contracting regulations. The best overall sup-

plier is then selected to participate on a long-term basis. Usually only one source is used, with both parties sharing cost information in order to focus on cost-reduction opportunities. Cost savings are shared by Thiokol and the supplier.

3. *Supplier certification.* A supplier model was developed after visits to companies demonstrating leadership in supplier base management. The model has five major elements: leadership, quality, delivery, cost, and technology. A supplier assessment program selected the candidate suppliers who would participate in certification. The assessment process is not a pass-or-fail system and evaluates suppliers on the five elements from the model. The questions are generic in nature to make the program universal and flexible. Open-ended questions are designed to encourage supplier input and participation.

Toyota

All suppliers must have a TQM program and be certified. Toyota gets involved in the suppliers' business right down to the manufacturing process. Toyota uses a system called *keirestu* that involves a network of suppliers with which Toyota has deep ties— and often financial links. These suppliers are privy to Toyota's secrets and are often willing to sacrifice some profits for the parent's sake. Toyota has been accused of trying to create a keirestu in the United States.

Westinghouse

Westinghouse's small central corporate purchasing group is headed by Thomas V. Doyle, director of purchases. Doyle's group serves as a catalyst for the purchasing groups in the 15 self-sufficient business units. The mission of the group is to "provide leadership through projects aimed at improving the total quality and cost effectiveness of the sourcing process throughout the corporation." The ten people at corporate headquarters serve 880 management and professional purchasing people in 70 different locations. Westinghouse relies heavily on special-interest councils, such as the small disadvantaged business steering council, in this decentralized environment.

One key role the corporate group serves is to help individual units take advantage of the buying clout that Westinghouse has as a $5.2-billion purchaser of goods and services. Corporate purchasing uses commodity surveys to identify candidate products and services for national contracts. Then commodity teams and Technical Advisory Committees (TACs) plan and execute contracts. Divisional representatives spend a day or two putting the strategy together, defining the requirements and specifications and identifying the potential bidders. Then engineering and manufacturing representatives are called in to form the TAC.

The corporate group issues a material price forecast quarterly to all the purchasing managers and controllers in the corporation. The forecast provides price projections on 47 commodities used at Westinghouse and an assessment of various commodities' vulnerability to supply interruptions from economic or political disruptions around the world.

Westinghouse's recipe for finding talent includes recruiting on college campuses for direct, entry-level purchasing positions in the field or for positions in their Purchasing Development Program (PDP), a six-month training program. At promotion time, the corporate group is used as a resource to identify people to fill openings.

Alan J. Meilinger, vice president for corporate services at the company, believes that purchasing will be a critical element in building the new Westinghouse as a result of 40 percent of the sales dollar being spent outside the company.

Xerox

Xerox, winner of the Malcolm Baldrige Award in 1989, achieved the following goals in dealing with its suppliers:

1. Reduction of its supplier base from 5000 to 350 worldwide, making for a concentration of economic leverage, more stable demands on suppliers, an ability to focus on a few high potential suppliers, and an ability to develop "model suppliers"

2. Use of a sole-source policy worldwide

3. Minimization of reliance on competitive bidding through the use of a target-cost method, while encouraging participation in design

4. Provision of extensive quality training for suppliers in Statistical Process Control (SPC), just-in-time purchasing, co-operative costing, set-up reductions, and Total Quality Management

5. Establishment of long-term (three-to-five-year) contracts

6. Adoption of a continuous supplier-improvement program that includes product design participation with open sharing of business and technical information

7. Formation of central commodity management teams

8. Use of competitive benchmarking for continuous measuring of processes, services, and practices against the toughest and best competition and functional leaders in the world

9. Establishment of multilevel communications with suppliers

As a result of these techniques, Xerox has realized an overall reduction in material costs of 50 percent. With improved designs, concentrated purchases, new technology, and internal efficiencies, Xerox has reduced defects from 10,000 parts per million (PPM) to 300 PPM.

Xerox representatives discuss "model suppliers" as they travel the country talking about how they won the Malcolm Baldrige Award. Model suppliers expect Xerox to be a model customer. The characteristics of a Xerox "Model Supplier" (MS) and "Model Customer" (MC) are as follows:

MS Company History: Shows organizational stability and demonstrated expertise

MS Financial Status: Has a stable position with growth potential

MS Management Attitude: Has customer satisfaction as a driving force; possesses a vision of the future; shows a willingness to change and progress

MC Management Attitude: Demonstrates honesty and avoids false promises; communicates openly about strategies and business constraints and expectations; provides a single-point business interface; is a significant customer; rewards performance with business growth

MS Quality: Products have defect levels in parts per million below the best in the industry. Embraces Total Quality Management concepts; practices statistical process control; is certified by Xerox and has its own certification program for its suppliers

MC Quality: Shares quality responsibility with suppliers; provides test specifications; develops universal quality metrics; provides competent subtier suppliers when required; communicates problems; fosters closed-loop corrective-action systems

MS Costs: Knows and controls cost elements and prices on cost, not market

MC Costs: Establishes realistic cost targets; provides benchmarking assistance; shares costs of change; provides prompt payment

MS Delivery: Demonstrates flexibility; delivers just-in-time, not just-too-late

MC Delivery: Strives for minimal schedule changes; develops realistic schedules and priorities

MS Service: Provides spares support and problem follow-up

MC Service: Provides training; shares design, commodity, and technical knowledge; provides simple and common communication systems

MS Technical Capabilities: Utilizes state-of-the-art production technologies; can process technical changes efficiently; has process, tool engineering, and prototype capabilities

Yale Materials Handling

Yale hosts and invites suppliers to participate in full-day seminars. One seminar is for executives, sales and manufacturing senior management, and quality supervisors. Another seminar is for the everyday doers, the frontline people. Yale has cut its supplier base from 1100 to 800 and is headed toward 300. Ideally, it wants two partner suppliers for each of the 91 separate commodities purchased.

Summary

Companies across the nation are implementing TQM purchasing activities to improve their relationship with their suppliers. Although each approach is different, a review of these companies is helpful in determining which approaches work best.

PART 2

A Model for Implementing TQM in Purchasing

5

Overview of the Implementation Model

Introduction

The second half of this book will present a seven-step model for implementing TQM in purchasing. The seven steps will be explained, one by one, in Chapters 6 through 12.

The specific model presented is a hybrid of the TQM model used by the Department of Defense (shown in Figure 5-1). The DoD identifies TQM as a focused management philosophy for providing leadership, training, and motivation to continuously improve an organization's management and operations.

The key element that distinguishes TQM from other improvement strategies is management's up-front commitment to its goals and strategies. If management is not wholeheartedly committed to this effort, don't even start. You will be wasting your time and resources.

The seven-step model we will be using is focused on meeting the needs of purchasing's customers—both requisitioners and suppliers. It uses a quality tool, Quality Function Deployment (QFD), to determine those customers' requirements. The model is

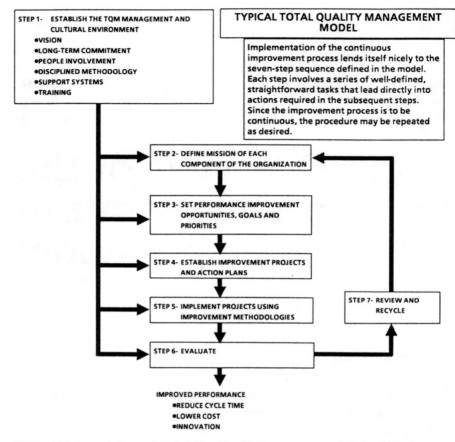

Figure 5-1. A typical Total Quality Management model. (*DoD Document #5000.51–G, as adapted from the Honeywell Aerospace and Defense Improvement Guide*)

only a guide and is to be applied in conjunction with traditional management approaches, such as quality awareness and employee involvement.

Figure 5-2 presents a graphic display of the basic TQM purchasing model.

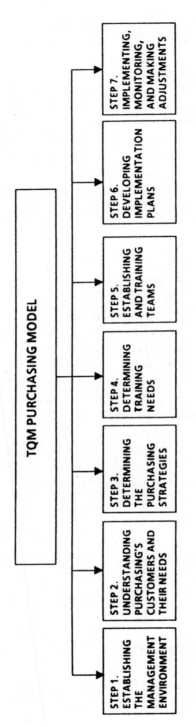

Figure 5-2. Total Quality Management purchasing model.

The Seven Basic Steps

Step 1: Establishing the Management Environment

The objective is to have management clearly demonstrate the leadership needed to establish the conditions for the process to flourish. Management's objective is to create a new, more flexible environment and culture that will encourage and accept change. This is not an easy task and will be the most frustrating for all involved.

Step 2: Understanding Purchasing's Customers and Their Needs

The objective is to genuinely understand purchasing's customers and their requirements. For the purpose of this model, purchasing has two sets of customers. The internal customer is the requisitioner for whom purchasing buys and to whom purchasing delivers products and services. The external customer is the supplier to whom purchasing gives information—such as drawings, specifications, dates, and so forth—that enables the supplier to provide products and services as needed.

A quality tool, Quality Function Deployment (QFD), is used to translate the customer requirements, known as Voice of the Customer (VOC), into purchasing strategies.

Step 3: Determining the Purchasing Strategies

The objective is to select purchasing strategies appropriate to the needs of the customers. Rarely is the first set of strategies selected successful. (Step 3 usually means kissing a lot of frogs before you find a prince.)

Step 4: Determining Training Needs

The objective is to determine the training needs of the purchasing department and of the supplier community.

Step 5: Establishing and Training Teams

The objective is the training of multifunction teams, some including suppliers, to serve as a vehicle for implementing the projects for continuous improvement. The structuring, makeup (leaders and members), and training of the teams are all important to success.

Step 6: Developing Implementation Plans

The objective is to identify tasks and develop implementation plans. Two types of plans are utilized: plans that will yield short-term results (three to six months), and plans that are more long-term (one year or more) in execution and results. (In addition, several TQM gurus will be discussed to facilitate an understanding of their approaches and views on implementation.)

Step 7: Implementing, Monitoring, and Making Adjustments

The objective is to establish a feedback loop that provides for monitoring the implementation of the plans and making adjustments to them based on the results.

6

Step 1: Establishing the Management Environment

Introduction

Establishing the Management Environment is Step 1 in purchasing's total quality management model. Figure 6-1 illustrates the steps that need to be taken to establish change within the purchasing group.

If your company is already involved in a process of continuous improvement, the establishment of the management environment may already be in place. However, if you are like most of us, the environment in your company is not yet conditioned to continuous improvement. As a result, implementing purchasing's role in TQM will require both an internal change process (within your organization) and an external change process (with your suppliers).

Management must create a new, more flexible environment

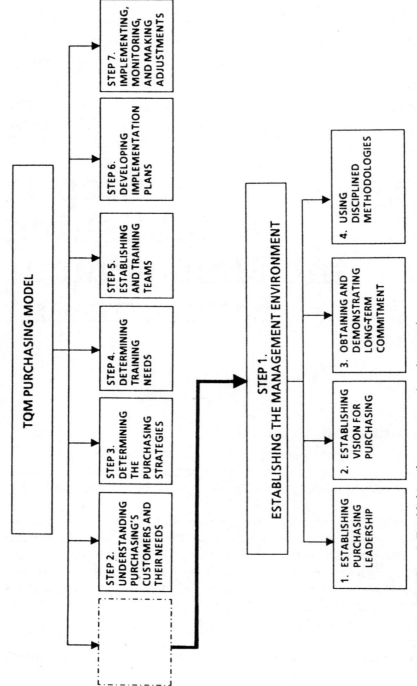

Figure 6-1. TQM Step 1: Establishing the management environment.

and culture that encourages and accepts change. Examples of such a cultural change are shown in the table below.

Category	Previous State	New Culture
Customer requirements	Incomplete or ambiguous understanding of customer requirements	Use of a systematic approach to seek out, understand, and satisfy both internal and external customer requirements
Suppliers	Unidirectional relationship	Partnership
Measurement	Data-gathering oriented to problem identification	Data used to under-stand and continuously improve processes

In trying to begin the process of change, a director of procurement at ITT Defense, George Wojnicz, introduced a simple concept that directly impacted the way his purchasing staff viewed their suppliers. He insisted that the word *vendor* be replaced with the word *supplier* when referring to the companies that provided products and services. Wojnicz felt that the word *vendor* conjured up an image of those individuals who sell their products (hot dogs, umbrellas, etc.) on the street. He felt that the word *supplier* better described the companies that would eventually become his purchasing staff's partners. Throughout his correspondence he used a simple logo (see Figure 6-2). This action—direct and un-

Figure 6-2. It is not always necessary to use elaborate behavioral scientific methods to set the environment for change.

complicated—was very effective as a way of launching the change process.

It is not the intent of this book to explain all the activities needed to create the correct environment for the implementation of TQM in purchasing. The process of change is not easy and can take a long time. Nevertheless, the purchasing manager must grasp the importance of change and, as far as possible, create an environment that is open to it.

Establishing Purchasing Leadership

Successful implementation of purchasing's role in TQM is a question of leadership. The purchasing management must assume the leadership role if the process is to be successful.

Many managers today do want to achieve TQM and have even announced a commitment to achieving it. But they tend to think that, having done so, they have done their part and that their subordinates will now pick up the ball and run with it. Each manager, however, must not only put TQM on the agenda, but put it at the very top. A leader must tirelessly pursue practices, procedures, systems, and people and head up efforts to develop reward systems that reinforce the new values. A leader must understand and accept certain basic truths:

- No one wants old-style bosses. (Even the bosses don't want old-style bosses.) The role of the leader is to direct from the top of the pyramid, making it possible for the front-line people to do their jobs.
- People want to do a good job. We should assume that rational, adult people will do what is asked of them and will also focus on those areas that get them rewards and recognition.
- Managers don't know everything. Certainly they don't know how to fix everything. Many managers don't know the details of all the work being performed by their subordinates. But while managers are not expected to know all the details, they should be able to provide the guidance their subordinates need to achieve satisfactory end results.

- Improving and sustaining performance is a continuous process and not a one-time effort.
- Before moving on to other departments, a leader must focus on cleaning up his or her own house. Managers should lead by example and not by words alone.
- Ownership of a process or product is the key to achieving quality. A spectator is less likely to do what is necessary to continue improvement than is the owner of the process under scrutiny.
- The first job of the leader is to demonstrate excitement about the future of the team's role in identifying areas for improvement.

Establishing a Purchasing Vision

One of the initial actions stipulated in the DoD model (Figure 5-1) is for management to create a vision of what the organization wants to be and where it wants to go. The vision should reflect the values of the organization and be unencumbered by current constraints and/or systems; creative thinking is critical. Examples of empowering visions include:

- President John F. Kennedy's vision of U.S. astronauts landing on the moon by the end of the 1960s
- Dr. Martin Luther King's vision of a better future for black Americans in his "I Have a Dream" speech

Companies actively involved in TQM spend long agonizing sessions developing vision statements. In many cases each word is intensely debated. Words such as "the best" or "a best" take on significantly different meanings and goals. The conclusion of the process should result in a management team's improved understanding and commitment to the vision it has articulated.

A vision is necessary for the purchasing department to begin its process. Whether you intend to be the best purchasing department in the world, develop the most supplier partnerships, or be the most cost-conscious is critical to setting the tone of your objective. Phrases such as "best in the world," "world-class supplier," "second to none" set the mind thinking about how to achieve these ideals. Only then can one begin the long process of attaining these levels.

Obtaining and Demonstrating Long-Term Commitment

Webster's dictionary gives several definitions of the word *commitment*, including this one: "an agreement or pledge to do something in the future, to carry something into action deliberately."

Commitment is like breakfast. Chickens are involved in breakfast (eggs). Pigs are committed to breakfast (bacon/ham). A leader puts his bacon on the table.

The senior management of your company must be *committed to*, not merely *involved in*, the TQM process. Since the purchasing role will include long-term relationships with suppliers and extensive requirements for training, the senior management of your company must be in it for the long haul.

The purchasing manager should demonstrate, through behavior and action, his or her commitment as the leader in the process. The message you get when someone says one thing and does another is that they either don't know what they are doing or can't be trusted. Likewise, when a leader makes statements or takes actions that do not support the company's goals, the message comes across loud and clear to the employees.

> Management is monkey see, monkey do, and you (the management) are the referenced monkey. As management style changes, so will the company.
>
> Dr. Ouchi, Author of *Theory Y*

Using Disciplined Methodologies

Purchasing management must use a disciplined approach to achieve continuous improvement. Persistent, disciplined application of continuous improvement methodology is a must.

Knowing what TQM is and knowing what tools and techniques are available are both necessary for success, but not sufficient for achieving it. No matter how good the systems and processes are, they will be of marginal value unless people use them in a disciplined manner. Having the discipline to work on TQM day after day so that it becomes a new way of life is the key factor for success.

The Lessons Learned

The lessons learned in the change process usually outnumber the successes. This is caused by the newness and difficulty of the process. Following are some "tips" from different companies.

- John Deere's experience indicates that the employees' response will go through three phases.

  Phase 1 is frustration. "For almost twenty years management has been asking us what's wrong. We've told them, over and over, and little or nothing has changed. And now, here we go again."

  Phase 2 is testing. If a company gets past Phase 1, the employees will test management. "Is management going to tell us everything?" "Is there a hidden agenda?" "Will management give us all the numbers?"

  Phase 3 is involvement. Only after Phases 1 and 2 have been completed will total involvement begin. At this point, rewards and measurements will begin to be the main points of discussion.

- During the process at least four types of people will be encountered and must be addressed by the organization.
  1. *Propellers.* The fast-track employees who will thrive on the process and be a catalyst for its success.
  2. *Anchors.* The people who will dig in and wait it out. By the heat of the summer they think, this flavor-of-the-month will have melted away.
  3. *Passengers.* The people who will put their seatbelts on and go along for the ride but have little influence or impact.
  4. *Torpedoes.* The people who will try to sabotage the process out of fear of change itself or fear of losing their jobs.

- The Douglas Corporation tried everything to improve communications with its staff—newspapers, videos, flyers. Although each of these methods was somewhat effective, they found the best way to communicate with their workers was by practicing MBWA (Management By Walking Around). Nothing works like face-to-face contact.

Summary

Without the commitment of top management, the establishment of the proper environment will not be successful, resulting in the failure of any TQM initiative. Purchasing executives must not only obtain senior management commitment, but also provide the proper purchasing leadership themselves.

7

Step 2: Understanding Purchasing's Customers and Their Needs

Introduction

There are many approaches to understanding customers and their requirements. One method that has been successful is Quality Function Deployment (QFD). QFD is a process that can assist a purchasing department to determine:

- What a customer's highest priorities are
- Why a customer may find purchasing's current performance level unacceptable
- How to accommodate a customer who wants purchasing to improve its performance by three levels

This information can greatly increase a purchasing department's ability to focus on areas that will not only improve performance but also ensure customer satisfaction.

The Definition of Quality Function Deployment

The American Supplier Institute defines QFD as:

> A system for translating consumer/customer requirements into appropriate company requirements at each stage, from research and product development, to engineering and manufacturing, to marketing/sales and distribution.

QFD was developed in Japan by Dr. Yoji Akao and was introduced to the U.S. by Don Clausing of the Massachusetts Institute of Technology. Although its primary focus was initially on iden-

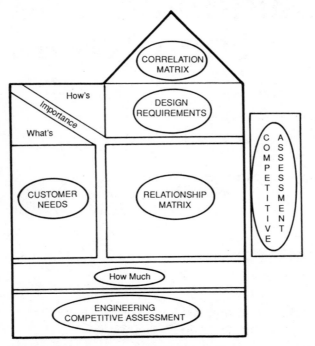

Figure 7-1. "The House of Quality"—the basic matrix structure. (*ITT*)

tifying customers' requirements for products, it is beginning to be applied to service-oriented processes like purchasing.

The QFD method uses a series of charts called "quality tables" to provide the discipline and communication required to focus on answering three questions: What? How? and How much? The quality tables are all part of a larger picture that is referred to as "The House of Quality" (see Figure 7-1).

From a purchasing perspective, QFD is viewed as a systematic means of ensuring that the demands of purchasing's customers are accurately translated into appropriate requirements and actions. QFD is a structured approach that provides:

1. A road map for identifying the customer's most critical requirements (known in QFD as the Voice of the Customer)
2. A starting point for selecting the most effective purchasing strategies to accomplish these goals.

Understanding Purchasing's Customers and Their Needs, Step 2 in the TQM purchasing model, is depicted in Figure 7-2.

Defining Purchasing's Customers

The use of QFD requires that the customer first be identified. As we have said previously, purchasing is a service function and has two sets of customers:

1. The internal customer is the requisitioner.
2. The external customer is the supplier.

The QFD Process for Purchasing

Applying the QFD process to determine appropriate strategies and actions for purchasing involves three phases:

Phase I   The first Quality Table (QT1) identifies the critical require-
                ents of the Voice of the Customer (VOC). Both purchasing
                and the customer are involved in this phase.

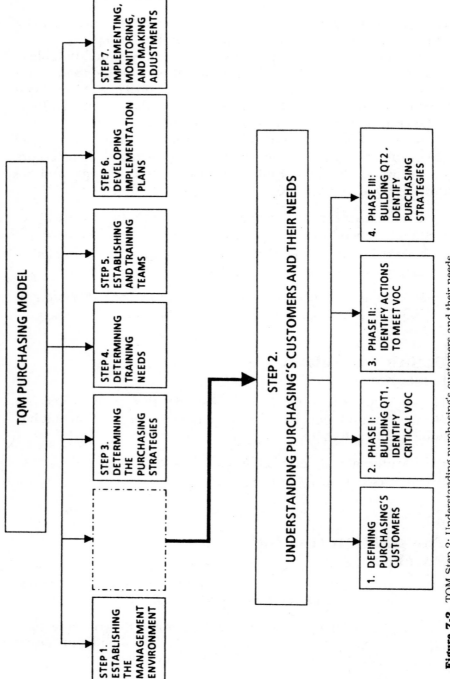

Figure 7-2. TQM Step 2: Understanding purchasing's customers and their needs.

Phase II   A series of specific actions are identified to meet the VOCs. Both purchasing and the customer are involved in this phase.

Phase III   A second Quality Table (QT2) identifies the most critical actions from Phase II. QT2 also identifies the purchasing strategies that will have the most positive effect on implementing the critical actions. Only purchasing is involved in this phase.

To the uninitiated a quality table that is used for purchasing may appear complex and difficult to understand. But in reality it is rather simple, as you will see as we build the tables.

Phase I: Identifying the Voice of the Customer

Six steps are required to complete Quality Table 1.

Step 1: The What, or Voice of the Customer (VOC), Is Identified.   The customer identifies what he needs from purchasing. This is known as the Voice of the Customer (VOC) or the *what* in QT1. Since purchasing has two customers, the needs of both must be identified. The VOCs are identified by answering the question: What does the customer expect of the purchasing department? It should go without saying that this information must come from the customer, not from purchasing's knowledge or ideas about the needs of the customer. (See Figure 7-3.)

Step 2: The VOC Is Prioritized.   The customer prioritizes each VOC, based on importance as determined by the customer. (Key: a full circle = Most important; an open circle = Moderate importance; a triangle = Least important.) (See Figure 7-4.)

Step 3: VOC Performance Is Evaluated by the Customer. The customer evaluates each VOC based on how he feels about the performance of purchasing in regard to it. (Scale: 5 = Good; 3 = Adequate; and 0 = Poor/Unacceptable.) (See Figure 7-5.)

Step 4: VOC Targets Are Established.   Purchasing and the customer jointly establish targets that will improve the customer's perception of purchasing's performance. (See Figure 7-6.)

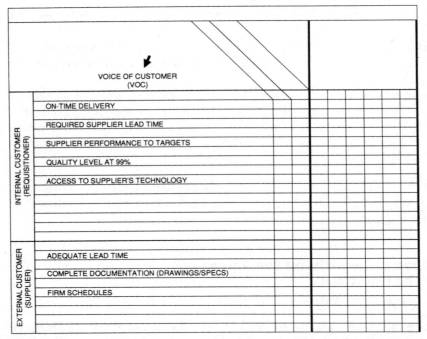

Figure 7-3. Quality Table 1, Step 1: The what, or VOC (Voice of the Customer), is identified.

VOICE OF CUSTOMER
(VOC)

INTERNAL CUSTOMER
(REQUISITIONER)

ON-TIME DELIVERY

REQUIRED SUPPLIER LEAD TIME

SUPPLIER PERFORMANCE TO TARGETS

QUALITY LEVEL AT 99%

ACCESS TO SUPPLIER'S TECHNOLOGY

EXTERNAL CUSTOMER
(SUPPLIER)

ADEQUATE LEAD TIME

COMPLETE DOCUMENTATION (DRAWINGS/SPECS)

FIRM SCHEDULES

IMPORTANCE
● MOST IMPORTANT
○ MODERATE IMPORTANCE
△ LEAST IMPORTANT

IMPORTANCE

Figure 7-4. Quality Table 1, Step 2: The VOC is prioritized.

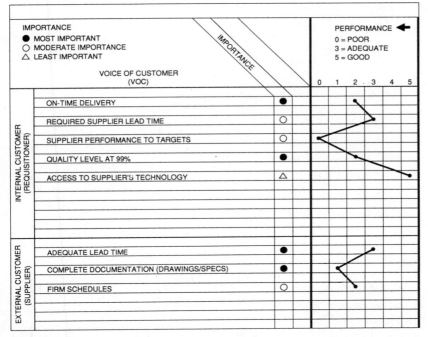

Figure 7-5. Quality Table 1, Step 3: VOC performance is evaluated by the customer.

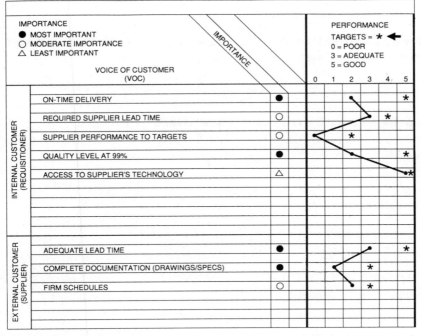

Figure 7-6. Quality Table 1, Step 4: VOC targets are established.

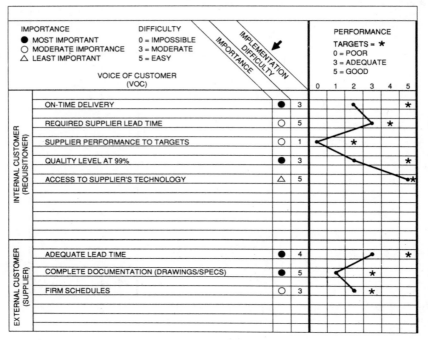

Figure 7-7. Quality Table 1, Step 5: Each VOC's degree of implementation difficulty is determined.

Step 5: Each VOC's Degree of Implementation Difficulty Is Determined.

Purchasing determines how hard it will be to implement each VOC. (Scale: 0 = Impossible; 3 = Moderate; 5 = Easy.) (See Figure 7-7.)

Step 6: Critical VOCs Are Identified.

Purchasing and the customer jointly identify the critical VOCs by analyzing importance, degree of difficulty, and customer perception of current performance. (See Figure 7-8.)

For example, the logic used in selecting one critical VOC—on-time delivery—might go as follows:

High customer priority = Strong importance

Performance level = 2 (Poor/Inadequate)

Implementation difficulty = 3 (Moderate)

Target improvement = 3 levels

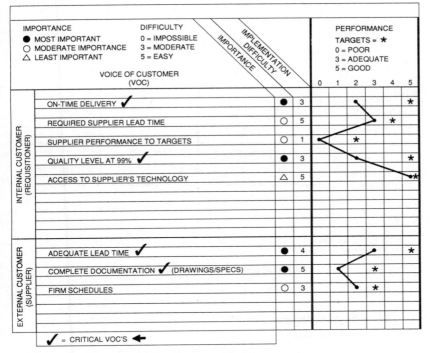

Figure 7-8. Quality Table 1, Step 6: Critical VOCs are identified.

Phase II: Identifying Actions to Satisfy the VOC

In this phase purchasing and the customer jointly identify actions—both those already in process and those still needed—to improve each critical VOC. Both internal actions (to be completed at the purchaser's organization) and external actions (involving the supplier) are identified. (See Figure 7-9.)

For example, in the case of the on-time-delivery VOC, the actions identified might include the following:

1. Expedite suppliers.
2. Visit suppliers.
3. Update computer system.

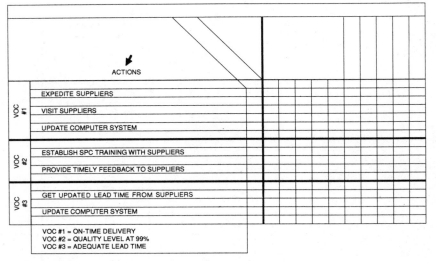

Figure 7-9. Phase II: Identifying actions to satisfy the VOC.

Phase III: Identifying Purchasing Strategies

Six steps are required to complete Quality Table 2.

Step 1: The Importance of Each Action to Its Own, or Primary, VOC Is Determined.   In the example, the action "Expedite Suppliers" is considered most important to improving VOC #1. (On-Time Delivery). (Strong = 3 points; Medium = 2 points; Weak = 1 point.) (See Figure 7-10.)

Step 2: The Impact of Each Action on the Other VOCs Is Examined.   Each action is then examined to determine if it has a favorable impact on the other VOCs. (For each favorable impact, the action is given 1 point.) In the example, the action "Visit Suppliers" will help improve not only VOC #1, but also VOC #2 (Quality Level at 99%). (See Figure 7-11.)

Step 3: The Total Impact of Each Action Is Used to Determine the Most Critical Actions Necessary.   The total of steps 1 and 2 (in Quality Table 2) determines the most critical actions necessary. The priority of these actions (I, II, III) is determined using the unnumbered table on page 88. (See Figure 7-12.)

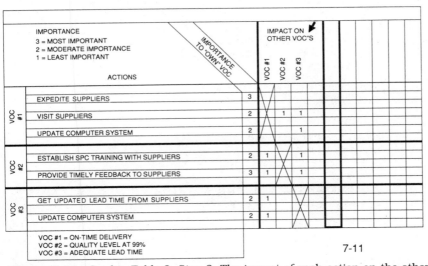

Figure 7-10. Quality Table 2, Step 1: The importance of each action to its own, or primary, VOC is determined.

Figure 7-11. Quality Table 2, Step 2: The impact of each action on the other VOCs is examined.

	ACTIONS	IMPORTANCE 3 = MOST IMPORTANT 2 = MODERATE IMPORTANCE 1 = LEAST IMPORTANT	IMPORTANCE TO 'OWN' VOC	VOC #1	VOC #2	VOC #3	TOTAL	ACTION PRIORITY				
VOC #1	EXPEDITE SUPPLIERS		3				3	II				
	VISIT SUPPLIERS		2		1	1	4	III				
	UPDATE COMPUTER SYSTEM		2			1	3					
VOC #2	ESTABLISH SPC TRAINING WITH SUPPLIERS		2	1		1	4	III				
	PROVIDE TIMELY FEEDBACK TO SUPPLIERS		3	1		1	5	I				
VOC #3	GET UPDATED LEAD TIME FROM SUPPLIERS		2	1			3					
	UPDATE COMPUTER SYSTEM		2	1			3					

VOC #1 = ON-TIME DELIVERY
VOC #2 = QUALITY LEVEL AT 99%
VOC #3 = ADEQUATE LEAD TIME

Figure 7-12. Quality Table 2, Step 3: The total impact of each action is used to determine the most critical actions necessary.

Priority	Action	Importance Own	Total VOC
I	Actions most important to achieving own VOC and having highest impact on other VOCs	3	4–7
II	Actions most important to achieving own VOC only	3	N/A
III	Actions not important to achieving own VOC but having a high impact on other VOCs	1–2	4–7

Step 4: Potential Purchasing Strategies Are Identified.
Potential strategies (e.g., partnerships, reduction of supplier base, etc.) are identified and examined. (See Chapter 8 for a discussion of strategies.)

Step 5: The Impact of Each Potential Strategy Is Compared to Each Critical Action. The object is to determine how each strategy will best assist in the implementation of each action. (See Figure 7-13.)

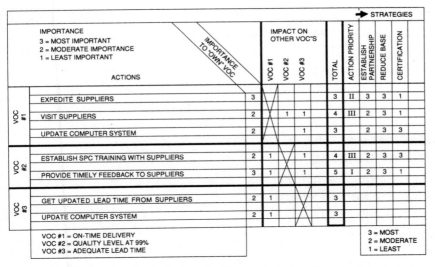

Figure 7-13. Quality Table 2, Step 5: The impact of each potential strategy is compared to each critical action.

Step 6: Appropriate Purchasing Strategies Are Selected.

Based on the highest total impacts, appropriate purchasing strategies are selected. (See Figure 7-14.)

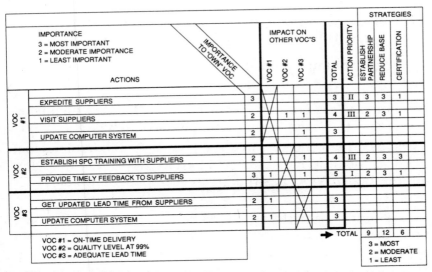

Figure 7-14. Quality Table 2, Step 6: Appropriate purchasing strategies are selected.

The above process provides a starting point for designing purchasing's role in TQM. The process may have to be repeated several times to select the most appropriate set of strategies. Subsequent chapters provide information to help managers transform this logic into action.

Alternative Use of QFD for Purchasing

An alternative use of QFD is to prioritize actions relating to VOCs without regard to selecting appropriate purchasing strategies. Phase I (Quality Table 1) and Phase II (identifying actions) remain the same. In phase III (Quality Table 2), steps 1 and 2 remain the same. A new step 3 (below) replaces previous steps 3, 4, 5, and 6.

Step 3: The Actions to Be Taken Are Prioritized. The following criteria are used to determine priorities:

1. The importance of the VOC to the customer
2. The performance of each VOC as determined by the customer using the following scoring: 1–2 = Poor; 3 = Average; 4–5 = Good
3. Each VOC's degree of implementation difficulty, scored as follows: 1–2 = Difficult; 3 = Moderate; 4–5 = Easy

The sequence of implementation is based on the following hierarchy:

First priority: If performance is poor (1–2) and implementation is easy (4–5), implement action immediately.

Second priority: If performance is poor (1–2) and implementation is difficult to moderate (1–3), establish priority by determining which VOC has the greatest impact on other VOCs (Step 2).

Third priority: If performance is average (3) and implementation is easy (4–5), use subjective analysis to determine whether action is appropriate.

Fourth priority: If performance is average (3) and implementation is difficult to moderate (1–3), establish priority by determining which VOC has the greatest impact on other VOCs.

Fifth priority: If performance is good (4–5) and implementation is difficult to moderate (1–3), establish priority by determining which VOC has the greatest impact on other VOCs.

Sixth priority: If performance is good (4–5) and implementation is easy (4–5) no action is required.

Lessons Learned in QFD Application

There is no set time to accomplish this process. It took four two-hour sessions just to define the process. As this technique is used, it is important to devote sufficient time at the front end to understanding the process and adequately documenting all areas.

One of the most critical lessons learned in the application of QFD is how much time is usually wasted in trying to understand what happened in a previous stage. Specifically, as each stage is completed, detailed documentation is required to include an explanation of each term used. (We constantly referred to our dictionary to avoid using the same terms over and over. The dictionary also gave us the opportunity to refresh our memories as to what we meant by each term.)

Summary

The use of QFD to determine the needs of the customer provides a baseline and logic for the selection of appropriate purchasing strategies. The process will have to be repeated over time as customer needs change.

8

Step 3: Determining the Purchasing Strategies

Step 1: *Establishing the Management Environment,* and Step 2: *Understanding Purchasing's Customers and Their Needs,* focus on deciding where the purchasing organization wants to go. Step 3: *Determining the Purchasing Strategies,* identifies the tactics, methods, and approaches needed to get there. (See Figure 8-1.)

Introduction

In TQM the use of several purchasing strategies may be required. The selection of a set of strategies will change as the organization becomes familiar with them and gets feedback from short-term implementation plans. Nine strategies that have proved particularly successful are:

1. Developing strategic relationships/partnerships
2. Reducing the supplier base and/or relying on sole sources
3. Implementing supplier certification

4. Practicing Just-In-Time (JIT) purchasing
5. Reducing variability
6. Improving cash management
7. Benchmarking
8. Involving suppliers in design (concurrent engineering)
9. Going back to basics

Strategy selection is not an easy process. Although many of the leaders mentioned in Chapter 4 indicate that their strategies were successful, the chances of picking a winning set of strategies first time out are fairly slim.

During one of our initial attempts at selecting a set of strategies, we found ourselves unable to decide on any specific plan. So we decided that our initial objective would be to let a strategy evolve. Meanwhile, we implemented a variety of pilot programs to see which one was best for us. We then applied the most successful programs to high-impact areas, with the idea of applying them to all areas in future.

(A word of caution: When I presented this plan to our top management, I received several negative responses. How could I put a plan together without knowing where we were going? How would we know when—and if—we got there? Well, I had expected this reaction and asked top management if *they* knew what should be done. When it became clear that they didn't, they soon understood and accepted our approach.)

A discussion of the nine successful purchasing strategies follows to assist purchasing organizations in making their selections.

1. Developing Strategic Relationships/Partnerships

A Definition of Strategic Partnerships

In 1984, *Business Week* stated that, "For companies large and small, collaboration is the key to survival." Collaboration takes many forms:

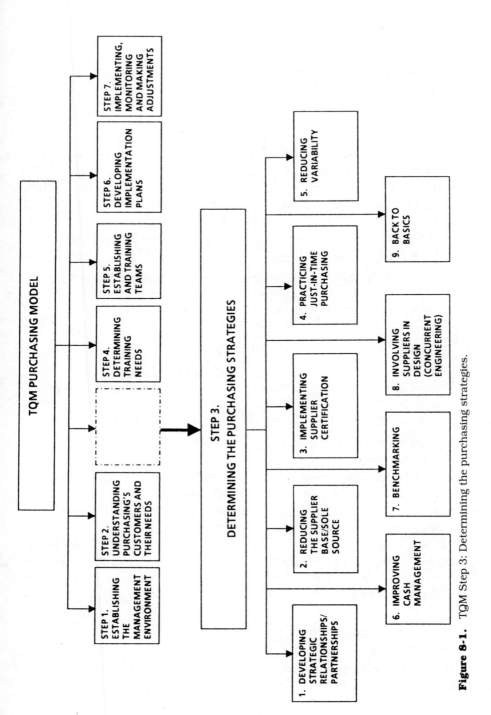

Figure 8-1. TQM Step 3: Determining the purchasing strategies.

Joint ventures

Technology exchanges

Equity positions

Long-term supplier partnerships

Strategic relationships with suppliers take the form of long-term supplier partnerships. Partnerships are viewed both as a strategy and as a result of implementing other strategies. The latter view assumes that, in order to achieve any type of improvement with suppliers, some form of partnership is required.

Supplier partnerships are not long-term contracts with a 30-day cancellation clause buried in the terms and conditions. Supplier partnerships are defined as:

> A mutual, ongoing relationship between a buying firm and a supplying firm, involving a commitment over an extended time period, and entailing a sharing of information as well as a sharing of the risks and rewards of the relationship.

Corning, Inc. uses the following definition:

> A partnership between a buyer and seller is a special, rare oc-currence which develops over a period of years between com-panies who trust each other, have common values, and are in-terested in each other's success.

For some, partnering conjures up visions of battalion-sized com-mittees and twelve-pound implementation documents. But part-nering isn't all rules and regulations. It's a state of mind, an out-look. Partnering encourages companies to share marketing, research, and production objectives, to build on compatible strengths and values, and to share risks and profits.

Effective partnering isn't a quick-and-easy fix that will rocket a company into the profit mode. It's a long-term working relation-ship built on trust. (An old Chinese proverb states that one gen-eration plants the trees and the next generation sits under them.)

Strong Versus Weak Partnerships

As Craig Stainbrook pointed out in a 1990 issue of *Electronic Buyers News*, supplier partnerships can be compared to marriage in America. If you marry for the wrong reasons, such as "everybody else is getting married," or "I want to have a lot of kids," or "I'll be too old if I wait any longer," the marriage may be destined for divorce.

Likewise, if a long-term supplier partnership is set up for the wrong reasons, it too may be headed for divorce. If the answer to any of the questions below is "no," the partnership lacks the requirements for permanency and is probably doomed from the start.

1. Do both parties share equally in the risks and tangible benefits of the relationship?
2. Are all costs, pricing, profit, and technical information shared by both parties?
3. Are both companies' top management involved in the partnering process?
4. Do the users of what is bought support the partnership?
5. Does the agreement say or imply "till death do us part"?

Careful partnering avoids poor relationships. A company may have good supplier relationships and not have partnerships. Conversely, a company may have poor supplier relations, yet have partnerships with several key suppliers. The answer lies in treating the person who does not receive the order as promptly, courteously, and fairly as the person who does receive the order. These actions build good supplier relations. All companies can do with good supplier relations, since that translates into good customer relations.

Partnership Benefits
A lot has been written on the benefits of longer-term supplier partnerships. Some of these benefits include:

- Promoting a sound relationship that encourages both parties to make decisions in favor of the relationship
- Enhancing an investment of resources aimed at improving quality and reducing total cost
- Developing a common language and understanding between both parties

- Encouraging both parties to work together in several areas:

  Operational decisions
  Causes of variation
  Improvements in the customer's products
  New-product introductions
  Handling of inventory
  Eliminating "conference room decisions"

From the supplier's perspective, one of the most significant advantages is the opportunity to "market the relationship." Being able to indicate that a supplier is a preferred or a certified supplier of a Ford or a Motorola gives the supplier additional leverage in selling his product. Purchasers must ensure that their lawyers will allow the supplier to do this.

The potential results of this type of relationship are significant because they

- Allow the supplier to take risks, such as investing in:

  technology (process automation)
  equipment (process monitoring)
  training and facilities

- Provide the purchaser with the means to forecast the quality, cost, and availability of future material
- Provide for automatic price reductions based on continuous process improvement
- Eliminate buffer stock and allow for a move to Just-In-Time inventory systems

Supplier partnerships are not new. In the 1970s many of us developed partnerships because of a potential shortage of basic commodities (e.g., steel and copper) during a period of high demand.

In the early 1970s I was involved in developing purchasing strategies for a large corporation. We selected three main steel mills and developed long-term supplier partnerships. Our guarantee was that we would buy a specified percentage of our steel requirements over several years. In exchange, we wanted to be considered a preferred customer in times of shortage. To compensate for changes in demand, we opened our books to the steel mills so they could see how we forecasted. The process was effective and a steady supply of steel became available.

Actions to develop partnerships vary, but in most cases they require purchasing to make the investment up front, at the design stage of a product. Examples of some companies' partnership strategies include the following:

- *Intel, Inc.*

  Sets up "executive partners" at a high level. The executives from both parties act as the "godfather" of the relationship and ensure that it is sound at all levels.

  Allows a supplier's personnel to work at Intel for up to one year and then return to the supplier.

- *Federal Express*

  Recommends that partnerships be developed with internal customers first, so that purchasing understands the requirements of a relationship before trying partnerships outside.

  Has partnerships with several catalogue houses; customers calling the catalogue houses are actually talking to Federal Express employees who take and process orders.

- *Honda.* Develops partners for life.

- *Douglas Aircraft.* Sets up partnerships for the life of the product.

- *Textronic, Inc.* Never bids for the two hundred preferred suppliers that are not at risk for losing their business.

The most successful illustration of partnering I have discovered involves SEMATECH (Semiconductor Manufacturing Technology). SEMATECH was incorporated in 1987 and is in partnership with 14 U.S. manufacturers of semiconductors. With the threat of losing an entire industry to foreign competition, this group of companies, in cooperation with the U.S. federal government, formed SEMATECH. Its mission is to provide the U.S. semiconductor industry with the domestic capability for world leadership in manufacturing. SEMATECH and its 130 supplier partners agreed to work together to ensure the survival of their industry.

Starting a Partnership

In a 1986 *Purchasing* magazine article, Walter Goddard, of Oliver Wight Education Associates, suggests that a buying company interested in establishing a partnership with a supplier should set up a meeting with the supplier. This meeting, he says, should entail the ten steps listed below. The supplier's local sales representative and sales manager, as well as its executive management team, should attend the meeting—as, of course, should their counterparts at the company.

1. Welcome/Introduction. The CEO should set the tone for the supplier partnership meeting, clearly communicating the importance to both companies of building a new and improved relationship.

2. The Company History. A brief discussion of the company's origins and general history should be conducted. This should be the beginning of an "open kimono" sharing of information. The idea is that the more each knows about the other, the easier it will be to find areas of mutual benefit.

3. The Company Profile. The past and future plans of the company should be presented, using facts and statistics, and emphasizing areas of interest to the particular supplier.

4. The Company/Supplier History. The relationship of the supplier to the company should be presented in terms of participation over the years. The reason why the supplier's participation has been increasing or decreasing should be discussed.

5. The Current Planning System. The reasons for the recent change in company management practices should be explained. This should not be a lesson in computers, but a lesson in changing management philosophy.

6. A Statement of the Company's Needs. This should be a brief statement about what your company needs from the supplier: shorter lead times, cost containment, improved quality, etc. No commitment is being sought at this time. The objective is to establish a basis for the buyer to later negotiate a new contract. The buyer should be prepared to explain what his or her company plans to do to help the supplier.

7. The Supplier's Current Response to Company Needs. The supplier's current performance should be discussed with specific reference to its response to company needs. Extreme care should be taken to avoid creating a defensive atmosphere. It's better for the company to take the blame for a supplier's poor performance than to create ill will. It can be implied that the company itself is cleaning up its act, and that, where the supplier is concerned, the emphasis will be on future performance.

8. The Supplier's Comments.   At this stage the supplier can ask questions and discuss causes of past problems. The supplier should be given a chance to come back at another time if so desired. The supplier's team may be reluctant to discuss certain issues and may want to gather more information before commenting. The goal is to achieve better understanding, not to get a specific commitment.

9. A Tour of the Plant.   This is the moment for showing off the company facility and emphasizing the part of the operation where the supplier's product can be used.

10. Lunch.   To prove the company's intent to improve the relationship, the company should host the lunch. Company representatives should continue to emphasize, by word and action, that the future relationship will be a sharing, open, and mutually beneficial experience.

Partnership Concerns

If partnering were an easy task, everyone would be doing it. There are many issues of concern to one or both parties in setting up a partnership.

Can the supplier have partnerships with competitors?
Yes, but the supplier should not share proprietary information.

How do you ensure that the information shared does not get into competitors' hands?
The legal answer is to sign a nondisclosure statement. The realistic answer is that there is no way to ensure this. A philosophy of trust will have to prevail. (It can be surprising to learn that many "secrets" really aren't that secret anyway. In many industries the speed of change in technology voids so-called secrets.)

Will the lawyers accept partnering?
One large hurdle *is* the lawyers, who can destroy any good partnership by requiring that everything be in writing. The word *trust* in a contract can soften up the cancellation clauses. This sends most lawyers up a wall.

What do you do if your partner's competitor comes up with a technology breakthrough?
It is the obligation of the supplier to keep abreast of his competition and stay competitive. The buying partner is obligated to help the supplying partner achieve this competitiveness. But if the two cannot overcome the breakthrough, the partnership may have to be dissolved or modified. However, it is appropriate to ensure that both parties understand the situation, and it is desirable to have an agreement to end or modify the partnership under certain circumstances.

Are written contracts required?
An argument can be developed on both sides of this issue. Written requirements clarify understandings and eliminate potential confusion. However, having to write everything down undermines the basic philosophy of trust. The only appropriate answer is, it depends on the situation. I believe it is better to write down all critical elements of the agreement to avoid confusion. This will also help eliminate the lawyer roadblock.

How long does it take to develop a true partnership?
There is no set answer, except that it takes time. The Harris Corporation has been working on partnering since 1985 and has three strategic partners. Douglas Aircraft claims eleven partners after one year of effort.

Partnership Lessons Learned

- One of the biggest mistakes a purchasing department can make is to institute supplier partnering by itself. Purchasing needs the support and cooperation of quality, distribution, marketing, accounting, production, engineering, and top management. When management is committed to long-term planning, partnering can work.

- Purchasing decisions that affect partnerships and profit are made at the desk of the buyer. Surveys show that only one out of three buyers goes out on plant visits. The buyer needs to understand the suppliers in making decisions. The best way to do this is to go out and meet the suppliers.

- Intel, Inc. warns against having too many partners. This could lead to a "me-too" program. The thrust should be to restrict partnerships to those that are really needed.

- Partnerships require educated procurement professionals. The establishment of long-term relationships will require the buyers to know more about the partners than their product capabilities. They must also understand the financial picture and have some technical expertise to help in mutual problem solving. Textronics requires all first-level buyers to have a two-year degree and all others, a four-year degree.

- Avoid falling into the Japanese keiretsu trap. Most U.S. companies have never heard of keiretsu but many, especially auto companies, may be among its victims. Keiretsu is the system that benefits Japanese auto makers at the expense of parts makers. The system was revealed by an anonymous Japanese auto executive in a recent congressional testimony. The executive indicated that the system is being established in the U.S. as Japanese transplants increase their operations. The injustices of keiretsu include under-the-table kickbacks and forced price reductions that make it difficult, if not impossible, for parts makers to turn a profit. The keiretsu is a cartel comprised of a dominant Japanese manufacturer and several suppliers. Each of the group's members owns part of the others, but the manufacturer is the controlling force. Since each member of the keiretsu deals exclusively with the others, American companies are blocked from competition for the business. Keiretsu is not what partnering is all about.

2. Reducing the Supplier Base and/or Relying on Sole Sources

The reduction of the supplier base is advocated as a prerequisite for many of the other strategies, such as supplier partnerships and certification. The smaller base reduces variability in the parts purchased, allows for the focus on a few key suppliers for lasting partnerships, and reduces the administrative process of dealing with multiple suppliers.

Using multiple suppliers for the same part increases variability. This is considered by many to be the primary reason for reducing the number of suppliers. The ultimate goal for reduced variability is to have one supplier for each part. (Further discussion of variability appears later in this chapter.)

The Harris Corporation plans to reduce its supplier base by 25 percent annually over the next three to four years. Harris believes this is what will make its partnerships grow. Increasing the share of business awarded to high-quality suppliers will impact directly on Harris's cost and quality. The supplier will have an incentive to invest in better people and processes. Additionally, the reduced transaction load will give the purchasing and quality people more time to focus on the external factory (suppliers).

Reducing the supplier base was credited by one small company as the strategy that provided the resources to implement its purchasing TQM program. The reduction in administrative costs freed up funds to expand the program. Meanwhile, the reduction in suppliers immediately decreased the amount of variability in parts, thereby improving their quality.

It should be noted that some statistical surveys do not bear out the premise that reducing the supplier base leads to overall improvement. The results of a survey published in the January 1989 issue of *Purchasing* magazine, for instance, indicated that for some commodities, such as connectors and lubricants, buyers with multiple sources gave their suppliers a higher overall quality rating than those with single sources. It may be that the quality of these commodities has risen to the point where suppliers are competing in quality ahead of price and delivery.

I am more comfortable with the practice of "qualifying" the

supplier base. Reduction for reduction's sake is not always the right thing to do. In some situations, such as a new technology or the customer's inability to provide any type of forecasting of material demands, the expansion of a supplier base may be appropriate.

Those who advocate reducing the supplier base preach that the ultimate objective is to get one supplier for each part purchased. However, this can send a powerful distress signal to the buyer. Indeed, the words *sole source* send a chill up the spine of most purchasing people. As a buyer, I myself have been burned more than once by sole sources. Yet Dr. Deming, in Point 4 of his fourteen points of management, says, "End the practice of awarding business on price tag alone. Instead, minimize total cost. Move toward a single supplier for any one item in a long-term relationship of loyalty and trust."

Deming further states that there are "illusions" in multiple sourcing:

- If one supplier falters, there will always be a steady supply.
- Competition between suppliers will reduce prices and/or improve quality.
- Over time, eligible suppliers can meet specifications and supply sufficient output with little supplier-to-supplier variation.

Deming also indicates that there are significant advantages to single sources. They

- Reduce the variation coming into the customer's process
- Give the supplier greater volume and therefore greater opportunity to refine and improve the process
- Encourage and require more prompt and collaborative problem solving
- Allow frequent interactions with suppliers that are directed toward continuous improvement

Single sources can be a concern to the supplier. As a small manufacturer I was not comfortable with being any customer's single source. If I could not produce, for whatever reason, I put my customer in jeopardy, which could result in the customer eliminating me as a source after the shortage was resolved. In developing single sources, ask for the supplier's input. You may be surprised.

3. Implementing Supplier Certification

James H. Currier, assistant vice president for corporate purchasing at NCR Corporation, suggests that the missing link in many efforts to build close supplier networks is certification. He feels that certification and measurement should be used to "structure negotiations within a development plan to leverage joint resources" in reaching "mutual improvement goals." This, he says, "is significant to defining world-class supplier relationships."

The objective of supplier certification is to gain confidence, both in the information supplied by your company to the suppliers and in the suppliers' processes. This confidence has the ultimate goal of reducing dependency on incoming inspection by relying on suppliers' Statistical Process Control (SPC).

The benefits to purchasing are obvious. First of all, there is the financial benefit in not having to inspect parts. Purchasing spends 3 cents to 6 cents of each Purchasing dollar on inspection. On $5 million in purchases, the savings would be $150,000 to $300,000, and this is just a small part of the overall benefit. The potential reduction in cost due to elimination of rework is enormous.

What incentive does the supplier have for participating in a certification program? Some of the benefits include classifying the supplier in a favored- or preferred-supplier status. The knowledge the supplier gains from learning concepts such as Statistical Process Control can be applied to other processes. The supplier develops a level of understanding that allows him to pass along various certification requirements to his own suppliers.

Approaches to Certification

Approaches to certification vary. One purchasing group certified all suppliers at once with the idea that those who didn't provide quality parts would no longer be certified. Others have elaborate processes, having as many as 25 steps for each supplier to complete.

Certification is by part number or by commodity. For complicated assemblies the trend is to certify by part number. Raw materials are usually certified by commodity. The 3M company certifies by commodity and, once certified, allows the supplier to ship to various plants.

When our quality people wanted to pilot a certification program, they sought to minimize variation in the process. To that end they decided that for the pilot, only those suppliers who already provided good parts would be used. This would allow the quality people to focus on improving the process for certification and not worry about the parts.

Quality asked purchasing to verify the selection of suppliers. Purchasing had a fit. "That's great," they said. "You quality guys want to take a supplier that currently ships 99 percent good parts and get to 100 percent. Well, what about the problem suppliers we have—the ones that, for a variety of reasons, can't make any good parts? That's where we need help!"

A compromise was struck. Quality agreed to provide SPC training to the suppliers with problems while purchasing agreed to assist in the certification program.

Certification Issues

According to the American Society for Quality Control (ASQC) there are eight issues that need to be dealt with when establishing a supplier certification program:

1. Specifications
2. Site inspection
3. Supplier ratings
4. From ship-to-stock to Just-In-Time
5. Record keeping
6. Product and process monitoring
7. Communications and problem solving
8. Data evaluation

One of the most important aspects of these eight issues is supplier ratings. Suppliers are rated on a quantitative basis with measurements in quality, delivery, cost, and service. Subjective elements include cooperation, technical assistance, responsiveness to problems, and compliance with instructions.

ASQC lists several potential rating problems:

- The system is not documented.
- Reports are cluttered.
- The database management is inadequate.
- There is a lack of timeliness.
- There is a lack of supplier discrimination.
- There are operational problems.
- Management support is lacking.

The Downside to Certification

The downside to certification is that many small suppliers are being inundated with certification programs. Several key customers are simultaneously enforcing programs, each with different rules. The different approaches have become so widespread that many think a standard for certification is needed. Some believe that the Malcolm Baldrige criteria should be used.

Purchasing has at least three ways to go on this:

1. Certify suppliers to the Baldrige Quality Award criteria
2. Generate a standard of its own
3. Revert to playing hardball with multiple suppliers

Advocates and critics of the Malcolm Baldrige approach are already lining up.

> *Yes:* It is nonthreatening to a supplier. Companies know where they stand and can see their progress clearly.
>
> *No:* The Baldrige application is just a series of questions. It would be difficult, if you were a supplier, to achieve certification using these questions. Attempting to translate it into a standard, or a basis for certification, is an unwieldy task.

According to George Graham, senior vice president for corporate staff of Texas Instruments, an industry standard is not in order. Texas Instruments is both a supplier and a customer, with a U.S. supplier base of approximately 14,000. Graham says that the need for one certification process has never arisen. Although the different business entities are being called upon to respond to different certifications and different levels of requirements, these requirements are still the same when you get down to certification. If you set your goals high enough, you shouldn't have any problem meeting any certification requirement, states Graham.

Consider the perspective of a small supplier to Motorola. When Motorola required all its suppliers to apply for the Baldrige Award, several refused. One small midwestern distributor says that the decision was based on monetary considerations. The company was supplying about $25,000 worth of parts to one of Motorola's European units, a figure not large enough to justify the investment of time, money, and effort needed to fill out the Baldrige application. As the president of this company further stated: "My total management system is to buy product, inspect it, and ship it out on time. How do I refine that to make it better? Say `Cross my heart and hope to die that I'm shipping you good stuff'?"

Certification and Liability

If you certify a supplier, do you assume liability for defective parts? Lawyers will debate this question and can provide an acceptable set of reasons for either side. One concept is that the liability is shared by both parties and should at least be discussed in any arrangement.

Ask the Suppliers

When deciding to start a program, don't reinvent the wheel. Ask your suppliers for their input. They may already be participating in a program that will meet your needs. Remember, the objective is to gain confidence in order to eliminate inspection, not to put an additional burden on the suppliers.

4. Practicing Just-In-Time Purchasing

Just-In-Time (JIT) has many definitions. Some view JIT as purchases in small-lot sizes. Others have a broader view and define JIT as the reduction of cost through the elimination of waste. A. Ansari and B. Modarress, in their book, *Just-In-Time Purchasing*, believe JIT includes supplier selection and evaluation, bidding practices, incoming inspection procedures, inbound freight responsibilities, paperwork reduction, value analysis practices, and packaging aspects.

Regardless of the definition, the concept of JIT changes the way purchasing will perform. Following is a succinct summary of JIT's impact on purchasing.

Activity	Traditional Purchasing	Just-In-Time Purchasing
Selection of suppliers	Multiple-source	Single-source
Evaluation of suppliers	2 percent rejects OK	No rejects
Incoming inspection	Counting and inspection of a given quantity	Elimination of counting and inspection
Negotiation	Lowest possible price	Product quality with fair prices
Agreements	Short-term	Long-term
Paperwork	Formal paperwork a must	Less time on formal paperwork
Packaging	Regular packaging for every part number	Small standard containers hold exact quantities

SOURCE: Adapted from *Just-In-Time Purchasing*, by A. Ansari and B. Modarress (The Free Press, 1990).

Ollie Wight, an expert on materials management, draws the following analogy: "If manufacturing were a child, its first words would be 'Mama' and 'Papa.' The next words out of its mouth would be 'I'm out of parts.'"

Just-In-Time purchasing has the potential to eliminate "I'm out of parts" from manufacturing's vocabulary.

5. Reducing Variability

Focusing on reducing variability is a strategy that is becoming very popular. It is viewed as either the end product of another strategy (e.g., the result of reducing the supplier base) or as a distinct strategy all on its own. The latter approach focuses on reducing the variability of the parts received while ensuring consistency with the processes of the purchasing company.

The concept of reducing variability represents a dramatic change in the way parts are specified. The current practice—specifying ranges of acceptability for a product, with lower and higher specification limits and 100 percent conformance to the specifications—is becoming obsolete. The new trend is for specifying a target value (i.e., a specific parameter) and requiring maximum uniformity around the target value. Targets have nothing to do with specification limits. Figure 8-2 illustrates this concept.

Statistical Process Control:
A Tool for Variability

One of the tools used to control variability is the application of Statistical Process Control (SPC). SPC is a method for determining the cause of variation, based on a statistical analysis of the

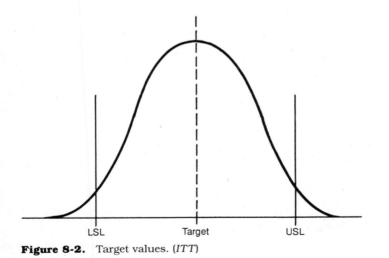

Figure 8-2. Target values. (*ITT*)

problem. Identifying problems quickly and accurately, SPC provides quantifiable data for analysis and promotes participation
and decision making by the people doing the job. Companies
deeply involved in SPC live by a code humorously summed up
as: "In God we trust: all others bring data."

Using a Variability Approach to Evaluate Suppliers

Under TQM, purchasing is switching from evaluating suppliers'
parts to evaluating suppliers' processes. It is another way of shift-
ing the focus from short-term to long-term considerations. If the
processes are under control, quality parts will be a natural outcome—not just for now, but over the long haul.

In evaluating a supplier's system for controlling the processes
involved in making a product, the following areas should be reviewed:

The method for identifying critical product characteristics

Repetitive failures responsible for high reject/cost rates (based
on Pareto analysis)

Dominant characteristics and process variables that significantly influence product quality

Product target dimensions or characteristics

Station or location (the step in the process) where measurement
will occur

Method of obtaining data

Reaction to out-of-control conditions—i.e., the action that will
take place in the event of process control

The Automobile Industry and Variability

The automotive industry provides a good example of the impact
of reducing variability. One of the key positions on the assembly
line is that of the door hanger. This is a highly skilled job that is
actually more of an art than a science. The door hanger must en-

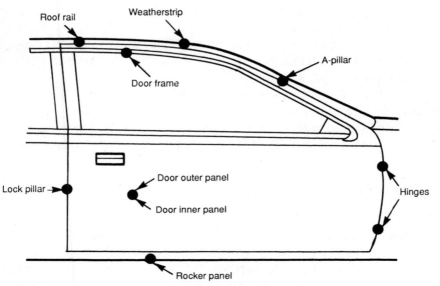

Figure 8-3. Variables influencing door fit. (*ITT*)

sure that a car's doors do not leak, that they line up with the other parts of the car, and that they shut with the distinctively robust sound of a secure closing. The American door hanger uses very sophisticated tools: a rubber mallet and a block of wood. Figure 8-3 illustrates the variables influencing door fit.

In studying Japanese automobile companies, Americans have been keenly concerned with how the Japanese fit their doors. Initially, when they were asked about this process, the Japanese were confused. They did not understand what the Americans were talking about. The Americans learned that the Japanese do not have door hangers. They simply put the doors on the hinges and the doors fit. The Japanese have eliminated variability in the key parameters.

6. Improving Cash Management

One technique that is not often talked about is improving cash management. The golden rule of cash is: THE ONLY IRREPARABLE MIS-

TAKE IN BUSINESS IS TO RUN OUT OF CASH. Almost any other mistake in business can be remedied in one way or another but, WHEN YOU RUN OUT OF CASH, THEY TAKE YOU OUT OF THE GAME! This rule applies to companies of all sizes. The recent trouble Donald Trump has had with his holdings is an illustration.

One of the easiest ways to improve relationships with suppliers and have them consider you a favored customer is to pay on time. This is especially true for small suppliers. Ask your controller to implement a policy that enables you to do the following:

- Pay all small suppliers in ten days.
- Override the computer system when necessary.

As the CEO of a small manufacturing company, my production department had one rule it was not allowed to violate: to be sure that my "top ten" customers (whom I had personally designated) had their product shipped ahead of everyone else. These were the customers who paid in ten days. Even though the production people would be pressured by other customers, my "top ten" always received priority. I needed the cash to run my business.

To save money, a common approach used today is to pay in 30 to 45 days but take all discounts. Reimbursement for unearned discounts is sent to those suppliers who request it. This is a practice that most large corporations use. (As a small businessman I understood this well. The first thing I did was add the discount to the unit price. I was reluctant to ask my customers for reimbursement of unearned discounts. I feared it would hurt my relationship with them.)

7. Benchmarking

A benchmark is a standard of excellence or achievement against which other similar things must be measured or judged. Benchmarking compares a company's performance against the best in the industry (its direct competitors), against companies recognized for superiority in performing certain functions (the "best in class" or world class), or against companies that are just better. Benchmarking establishes how much a company needs to improve to be at world-class levels of functionality. Benchmarking also:

- Provides identification of best practice(s) in any (all) process(es)
- Provides a quantitative basis for setting process improvement targets
- Encourages participating managers to set lofty goals

Benchmarking can be very revealing. According to the *McKinsey Quarterly* (1991, Number 1):

- A U.S. manufacturer of electronic circuit card assemblies discovered through benchmarking that its total production costs were four times higher, its quality performance one-fifth as good, and its total time expended on nonproductive activities twice that of one of the world's best producers.
- A pharmaceutical company found through benchmarking that its purity standards were one-tenth as stringent as those of a potential competitor.

Unfortunately, most companies have little information about who or what is the best. Consequently, most planned improvement targets are set internally, based on past performance. The *McKinsey* report indicates the following reasons for the little progress in benchmarking:

- Given a specific problem, our engineers are taught to break it down into simple concepts and first principles, rather than look at others' experiences for answers.
- There is a widespread belief that we are unique, especially when confronted by superior practices.

But, behold, help is on the way.

APQC Benchmarking Clearinghouse

In June 1991, the American Productivity and Quality Center (APQC) in Houston, Texas, created the APQC Benchmarking Clearinghouse to assist commercial firms, nonprofit organizations, and government departments in the process of benchmarking. The clearinghouse is intended to be the leading source on national and international information about "best practices" through databases, case studies, publications, seminars, conferences, videos, and other resources. It will also provide training and consulting on benchmarking methods. The 50 organizations that joined the APQC became operational in the first quarter of 1992.

Who Are the Benchmark Companies?

The identification of benchmark companies will change as the criteria for "the best" continues to improve. The following are some of the companies or agencies currently considered benchmarks in the areas indicated.

Area	Company
Assembly automation	Ford Motor Co.
Benchmarking	Alcoa
	AT&T
	Digital Equipment
	Florida Power and Light (only U.S. Deming-Prize winner)
	Ford Motor Co.
	IBM Rochester (1990 Baldrige winner)
	Motorola (1990 Baldrige winner)
	Xerox (1989 Baldrige winner)
Billing and collections	American Express
Concurrent engineering	NCR
	Motorola
Customer focus	Wallace Co. (1990 Baldrige winner)
	Westinghouse (1989 Baldrige winner)
	Xerox
Distribution	L.L. Bean
Document processing	City Corp
Employee suggestions	Dow Chemical
	Milliken & Co. (1989 Baldrige winner)
	Toyota
Empowerment	Milliken & Co.
Flexible manufacturing	Allen-Bradley
	Motorola
Industrial design	Black & Decker
	Herman Miller

Area	Company
Information systems	General Electric
Inventory control	American Hospital
Leadership	General Electric (Jack Walsh)
	Hanover Insurance (Bill O'Brien)
Marketing	Procter and Gamble
Quality (overall)	Florida Power and Light
Quality processes	IBM Rochester
	Florida Power and Light
	Wallace Co.
Research and development	AT&T
	Shell Oil
Self-directed work teams	Corning
Service parts	John Deere
Supply Management	Ford Motor Co.
	Motorola
	Xerox
Taguchi techniques	ITT
Training	Caterpillar
	Wallace Co.

NOTE: The Deming Prize, the highest quality award given by the Japanese, is the prize on which the Malcolm Baldrige Award was modeled. So far, only one American firm, a utility, Florida Power and Light, has been awarded the Deming Prize.

Xerox: A Benchmark of Benchmarking

The success of Xerox in the area of benchmarking is well known. In the late 1970s Xerox was faced with significant losses to its market share. It decided to compare the unit manufacturing cost and overall features of its copying machines with those of the competition. Xerox discovered that its competitors were selling units for the same price it was costing Xerox to manufacture its copiers. Loss of market share was, of course, a major concern to Xerox—especially to its senior management. But (as I learned at a training session given by one of Xerox's benchmarking managers) the galvanizing development was the loss of profit-sharing checks to the key managers. This got their attention, and the managers concluded that something had to be done.

Xerox used the findings about its competitors as improvement targets. The success of this approach soon made benchmarking a way of life at Xerox. Today, the company credits the technique with being one of the key elements that helped it win the Malcolm Baldrige Award in 1989.

Preparations for Benchmarking

Although companies have different approaches and processes in their benchmarking activities, most stress the importance of developing an environment conducive to benchmarking, preparing by doing homework before the process begins, and using the information gathered.

Finding Out How Much Need There Is for Benchmarking. Before starting benchmarking, Xerox determines whether or not it is called for. Benchmarking may be appropriate, for example, if there are no answers as yet to the following questions:

What factor is most critical to the business: customer satisfaction? inventory turns? expense-to-revenue ratio?

What areas are causing the most trouble?

What factors are responsible for customer satisfaction?

Where are competitive pressures being felt?

Benchmarking may also be appropriate when there is a need to expand the resources available to solve problems. (Given current restrictions on the amount of internal resources that can be applied to most problems, one option is to use benchmarking to access the expertise of other companies. Benchmarking can at least provide information about their solutions to common problems.)

Choosing "Better Than" Over "World-Class" and "Best-in-Class." Xerox does not seek to determine "world-class" and "best-in-class." Trying to benchmark the entire world or all the companies in an industry is not practical. Xerox's approach is simply to identify those companies that are "better than Xerox" and to concentrate on finding out how they achieved this level of performance.

Using Operating People Rather Than Staffers. Xerox's first attempt at benchmarking failed because it used staff people rather than the real "operating people" to do the benchmarking. One of the initial benchmarking efforts was done with Xerox's Japanese company, Xerox Fuji. The staffers visited Japan, but when they presented their results, were not believed by the operating people. Two years later the operating people themselves went to Japan and then became believers. (Among the things the operating people learned was that half the engineering was done by suppliers, and that Japan had long-term agreements with their suppliers.) Xerox emphasizes the use of the "owners of the processes" in conducting benchmarking. Once they become believers, operating people tend to start implementing changes immediately.

Committing Enough Time and People
According to Xerox, effective benchmarking requires a significant effort. Xerox's track record indicates that any benchmark completed in less than nine months has probably not been done adequately. Xerox has one hundred people working full-time on benchmarking, and all functional managers are expected to participate in repetitive updates of benchmarks and gaps in their own process.

A typical benchmarking activity takes six months, involves ten people, and benchmarks six to eight companies. The team consists of a facilitator, who usually serves as the leader (and can be a staffer) and two operating people for each of the companies to be studied. Some 60 percent of the leader's time is committed, with 10 percent from each operating member.

For a company's first attempt, Xerox recommends no more than four companies and nine team members (one facilitator and two operating people for each company to be studied).

**Xerox's Ten-Step
Benchmarking Process**

Xerox's process as described in "Competitive Benchmarking," put out by the Xerox Corporation in 1987, has five major phases that include ten steps. The real benchmarking activity is performed in Phases 1 and 2 (steps 1–5), with the remaining three phases (steps 6–10) emphasizing good management practices. (See Figure 8-4.)

Phase 1: Planning.   The objective of this phase is to lay out the plan for the benchmarking activity.

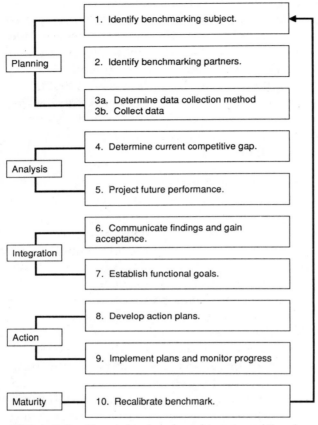

Figure 8-4.   Xerox's key benchmarking steps. (*Xerox*)

Step 1: Identifying What Is to Be Benchmarked.   The team selects an area. The area may be a product, service, process, or even a customer dissatisfaction area. The identification of what is to be benchmarked could come from the problem logs that are used in meetings or from a brainstorming session with senior management. Each area identified should have a key manager assigned as its champion (ideally, the senior manager who has most control over that area).

Step 2: Identifying Benchmarking Partners for Comparison.   The companies chosen can be competitors, other units within the company, or any firms that are considered leaders in their field. The identification of partners to participate in the benchmarking activity requires homework. (Information on prospective partners can be found in public libraries and financial reports and via survey groups, market research groups, and consultants.)

The benchmarking partners ultimately chosen should have characteristics similar to those of the firm doing the benchmarking—a company, for example, that has gained and lost market share in the last x years, or a company that has competition in the Far East, or a company that has the highest inventory turnover (if inventory turnover is the benchmark area), etc.

Even before the first partner is contacted, the efforts exerted in establishing a team, developing a questionnaire (see below), and selecting the benchmarking partners can take as much as two months out of a total six-month activity period.

Step 3: Determining Data Collection Method and Collecting Data. Once the area for comparison and the benchmarking partners have been identified, the next activity is the development of a questionnaire. *The questionnaire is key, because it becomes the basis for the balance of the activities.*

- *Creating the Questionnaire.* A quality tool that Xerox often uses to develop the questionnaire is the cause-and-effect diagram (also known as a fishbone, or Ishikawa, diagram). Data for the questionnaire can be collected via many avenues, including visits, person-to-person exchanges, and telephone conversations.

  The questionnaire should be designed so that "paragraph" responses are not allowed. Included should be schematics of processes, with blank sections next to each step for the partner to fill in, specific statistics with requests for similar statistics, and examples accompanied by requests for similar examples, etc. A list of definitions should also be included. If the list is longer than one page, it should be reduced, or the questionnaire itself changed.

- *Validating the Questionnaire.* On its completion, the questionnaire is answered by the team, with feedback provided to include the time it took and any general comments. Adjustments are made and then recirculated to be answered by other internal people. Two internal groups' answers to the questionnaire help to validate the questionnaire and provide the necessary information to determine what the company will actually do. The process also provides the company with information that can help determine whether additional training is needed in the process.

Xerox's rule of thumb is to ensure that the questionnaire takes no more than one hour to complete. If it is in sections, with parts that will be answered by different groups within the partner organization, each section should take no more than an hour to complete.

(When other companies ask Xerox to participate in benchmarking efforts, Xerox asks to be sent their questionnaires rather than just giving out information or providing statistics over the phone. A company's questionnaire acts as an indicator that it is serious and ensures that Xerox itself will be able to obtain some information by participating in the process.)

- *Contacting Partners.*   Initial contact with benchmarking partners is by telephone. The area to be benchmarked is explained, with emphasis on the sharing of information (exclusive of confidential information), the extent of the effort required of the partners, the meetings to be planned, etc. Xerox suggests that the initial contact be with each partner's vice president of quality, who is unlikely to turn down the request.

- *Sending Out the Questionnaire.*   The questionnaires should be sent out via overnight express, to give them an added air of urgency and importance. The next day, each partner should be called to be sure its package has been received.

  Allow partners a full month to answer the questionnaire, but either visit or, at a minimum, call each partner within two weeks to check on the questionnaire's status. (One of the objectives from this point on is to maintain the partners' interest in the benchmarking process. Be persistent, but be careful not to cross the line into harassment.)

- *Processing the Completed Questionnaires.*   Upon return of the questionnaires from the partners, Xerox immediately mails them its own team's answers to the questions to maintain their interest. The cover letter thanks each partner and asks that it review Xerox's responses and select five to ten items that its team would like to discuss further. Meanwhile, the Xerox team examines the partners' response and also selects five to ten areas. This process sets the agenda for the initial meeting, with those items that are mutually selected being given priority for the meeting. (There are times when a partner's responses do not generate any interest at Xerox, in which case, the partner is still invited to Xerox for the first meeting.)

- *First Meeting with Each Partner.* The initial meeting with each partner is usually at the partner's site. The Xerox team includes the facilitator and the two operating people assigned to that company. The partner is permitted to bring additional people, but the group stays together during the meetings.

Day one	Morning: Team flies to site.
	Afternoon: Team meets with partner.
	Evening: Team discusses results without partner.
Day two	Morning: Team meets with partner.
	Late afternoon: Team returns.

Xerox does not meet with the partner for dinner on the first day so that the team can confer. (Also, the expenses involved could be significant.) The return flight is scheduled for late evening of the second day to allow the team to discuss its results before returning to the home site.

- *Subsequent Meetings with Each Partner.* The second meeting is held at Xerox, with the same schedule and conditions as were followed for the first meeting. Each of the two operating team members and the partners are asked, prior to the second meeting, to sponsor and lead the discussion on one or two items. Each item should be given no more than a one-hour discussion. The Xerox team members serve as scribes when they are not leading the discussion. The facilitator's role is to ensure that all the essential questions get asked.

Phase 2: Analysis. This phase helps one to understand each partner's strengths and assess your own company's performance against those strengths. It determines the current gap, if any, and projects future performance levels.

Step 4: Determining Current Performance Levels. After the data is collected, the results are analyzed. Both positive and negative output are analyzed.

Step 5: Projecting Future Performance Levels. The team projects future levels of performance based on the data. This will assure that the target is pitched at the same level (or higher) as that of the benchmarked companies.

All partners receive copies of the questionnaire from all the participants. (The names of the companies are not revealed, and the results of the final analysis are not shared.)

Phase 3: Integration. This phase uses the data gathered to define the goals that must be attained to gain or maintain superiority. It includes the establishment of functional goals and the development of action plans.

Step 6: Communicating Benchmark Findings and Gaining Acceptance. The methodology, findings, and results are communicated, both to upper management and to the employees who will be asked to implement the actions for improvement.

Step 7: Establishing Functional Goals. The team presents ways in which the organization can improve and reach the goals established.

Phase 4: Action. During this phase the actions are implemented, monitored, and periodically assessed.

Step 8: Developing Action Plans. Action plans for each objective are developed, with a focus on assuring that the organization as a whole will accept them.

Step 9: Implementing Specific Actions and Monitoring Results. Action is initiated. Problem-solving teams are used if the actions do not show progress toward the planned goal.

Phase 5: Maturity. This phase is used to determine when the company has attained the leadership position and whether benchmarking has been a successful strategy in helping it attain that end.

Step 10: Recalibrating Benchmarks. Periodically the benchmarks are recalibrated to ensure that the goals are set on the latest performance.

AT&T's Twelve-Step Benchmarking Process

AT&T's twelve benchmarking steps, as described by Karen Bemowski in a 1991 article in *Quality Progress* magazine, are divided into two categories. The first six steps are referred to as the "first-things-first" steps because they set the stage for the benchmarking process by eliminating barriers. The next six steps comprise the real benchmarking process.

Step 1: Determining Who the Clients Are. Pinpoints who in the organization will use the benchmarking information.

Step 2: Helping the Clients Understand the Purpose of the Information. Assists clients to visualize what information will be gathered and how this information will be used as a goal to measure how to improve their performance.

Step 3: Testing the Environment. Involves spending time with the clients to understand their expectations and commitment to the process.

Step 4: Determining Degree of Urgency. Determines the priority given the process by the clients and how urgently they need the information. Those with new processes to develop tend to be the most receptive. Those with the desire to improve their performance will also be enthusiastic, but to a lesser degree.

Step 5: Determining the Scope and Type of Benchmarking Needed. Involves assessing the nature and extent of the resources needed (which tend to be proportional to the potential impact). The focus of the benchmarking determines what type of organizations will be benchmarked. AT&T benchmarks best-in-class, internal-best, and competitive-best.

Step 6: Selecting and Preparing the Team. Involves the selection of a six- to eight-member team. The team is responsible for the data collection plan and for the implementation of the actions. Generally only two to four members of the team visit the company being benchmarked.

Step 7: Verifying that the Benchmarking Process Is Part of the Business Planning Process. Involves meetings with top management to assure that the benchmarking activity is part of the overall business planning process.

Step 8: Developing Implementation Plan. Calls for the team to:

1. Prepare a mission statement that describes expectations
2. Prepare for data collection with assignments to members of the team
3. Develop a profile for selecting the benchmarking candidates
4. Do research and develop scripts to help manage the site visits
5. Answer the script questions with information about the clients' current performance (including the set of metrics now being used)
6. Invite the benchmarking partner to the initial meeting
7. Send the benchmarking partner an advance copy of the script to help its team prepare for the meeting

Step 9: Analyzing the Data. Compares the information obtained to the clients' current performance. To organize the data, flow charts and matrixes are used. The team identifies areas for improvement.

Step 10: Integrating the Recommended Actions. Integrates the recommended actions into the clients' planning and budget process.

Step 11: Taking Action. Implements the actions, using standard implementation procedures.

Step 12: Continuing Improvement. Ensures that benchmarking becomes a part of the continuous improvement activities. Benchmarks are periodically updated to ensure that the goals reflect the latest performance.

Alcoa's Six-Step Benchmarking Process*

Step 1: Deciding What to Benchmark. The project's sponsor determines what is to be benchmarked by answering the following questions:

- Is the topic important to the customer?
- Is the topic consistent with Alcoa's mission, values, and milestones?
- Does the topic reflect an important business need?
- Is the topic significant, in terms of costs or key financial indicators?
- Is the topic an area in which additional information could influence plans and actions?

The result of this step is a statement that sets the goal of the effort and outlines what is expected.

Step 2: Planning the Benchmark Project. The team is put in place. The leader (ideally the project sponsor) is chosen. The team refines the process by answering the following questions:

- Who are the customers for the study?
- What is the scope of the study?
- What characteristics will be measured?
- What information about the topic is readily available?

The team then submits a project proposal to the sponsor.

Step 3: Understanding Your Own Performance. The focus is on self-assessment. Included in this step are the selection of the critical areas that will be targeted and the creation of a baseline.

Step 4: Studying Others. The candidates for benchmarking are selected and the process of obtaining the appropriate data begins.

*As described in "Benchmarking: An Overview of Alcoa's Benchmarking Process," Aluminum Company of America, 1990.

Step 5: Learning from the Data. The data is analyzed and the gaps between the organization's performance and the data are identified.

Step 6: Using the Findings. The team works with the sponsor to implement activities and tries to determine what other parts of the organization can use the data.

Supply Management Is a Lucrative Benchmarking Area

Most benchmarking begins in the factory, with the visible capital or labor-intensive processes. But factory activities usually account for only a small percentage of the total costs, *making the factory the wrong place to begin benchmarking*. It makes better sense to benchmark those activities that account for the majority of the costs. Supply management is one of those activities. Knowing how competitive your suppliers are in the world market, how well you manage your suppliers, and how well you manage the cross-functional processes that promote excellent supply management will yield far more beneficial results than analyzing any process on the factory floor.

Approaches to Benchmarking Supply Management (Purchasing) Performance

In a 1990 *Purchasing* article, J. Cavinato, associate professor of logistics at Penn State University, states that there are eleven basic approaches to benchmarking:

1. Comparing total purchasing system performance against that of another company or division
2. Comparing total purchasing system performance against that of a competitor
3. Comparing total purchasing system performance against a quality operation in use by a firm in another industry
4. Comparing a specific purchasing component against that in another division

5. Comparing a specific purchasing component against that of a competitor
6. Comparing a specific purchasing component against that of anyone in the field
7. Comparing the firm's paid prices against those of the industry overall
8. Comparing actual budgets against planned budgets
9. Starting from scratch by comparing the results of a zero-based budgeting process against current levels
10. Baselining good practices in each area and comparing them to current levels
11. Asking internal customers what they think of purchasing's performance

Purchasing Performance Benchmarks

Benchmarking is not a new concept for purchasing. The Center for Advanced Purchasing Studies (CAPS), an affiliation agreement between the College of Business at Arizona State University and the National Association of Purchasing Management, has been involved in benchmarking for the last few years. CAPS was established in 1986 to address the needs of practical research in purchasing. CAPS has done research in such areas as organizational relationships, ethics, and purchasing from small minority-owned businesses.

CAPS current benchmarking efforts involves gathering purchasing performance data, summarizing the results as averages and ranges, and making this information available to participating firms and the general public. A future activity involves developing best-in-class information.

Over twenty-five industries are already complete or in process for the data-gathering piece. Among the many fields involved are the aerospace/defense, automotive, banking, computer, food service, petroleum, steel, telecommunications, and utilities industries. CAPS benchmarks 26 items, including total purchasing costs as compared with corporate sales and the percentage of cost savings generated by the purchasing department from cost reductions.

Benchmarking and the Malcolm Baldrige Award

The originators of the Malcolm Baldrige Award considered benchmarking a critical tool to assist companies in the improvement process. As a result, one of the seven Baldrige evaluation categories includes the use of benchmarking. In the 1992 award criteria, under *Category 2.0, Information and Analysis, Item 2.2, Competitive Comparisons and Benchmarks,* we find the following:

> Describe the company's approach to selecting data and information for competitive comparisons and world-class benchmarks to support quality and performance planning, evaluation, and improvement.

Those companies interested in applying for the Malcolm Baldrige Award can improve their chances by using benchmarking techniques. (See Appendix C for additional details on the Malcolm Baldrige Award.)

8. Involving Suppliers in Design (Concurrent Engineering)

"Our old focus on price yields puny results when compared with the 70 percent savings that have resulted from early supplier involvement in a concurrent engineering process." So says one chief executive.

Concurrent engineering is a new wave to an old concept. It is defined as a systematic approach to the integrated and overlapping design of products and their related processes, including design, manufacturing, and support. Concurrent engineering requires that, from the beginning, all elements of product life-cycle be evaluated across all design factors to include user requirements, quality, cost, and schedule.

The foundation of concurrent engineering is that some 80 to 85 percent of a product's cost is determined at concept development. Additionally, the integration of support processes early on cuts manufacturing costs while raising quality and reducing development time.

The significant benefits to be had from concurrent engineering include:

- Improved quality of design, leading to a 50 percent reduction in change orders
- A 40 to 60 percent reduction in product cycle time as a result of using concurrent, rather than sequential, design
- A 30 to 40 percent reduction in manufacturing costs as a result of using multifunction teams to integrate product and process
- A 75 percent reduction in scrap and rework as a result of product and process design optimization

Supplier involvement in this process is a necessary element for success. Since a company cannot be all things, the technologies and expertise of suppliers must be part of the entire process.

9. Back to Basics

For those of you who are overwhelmed with all these strategies (as I was when I first started to read about them), I offer an alternative approach. *Go back to the basics of good buying, which include understanding the requirements, understanding the abilities of the suppliers, and establishing good working relationships.* The outcome will be the use of the above strategies without the fancy terms.

Summary

Step 3 of the TQM model, *Determining Purchasing Strategies*, is one of the most critical and difficult steps to implement. Although a variety of successful strategies are available, the ability to select the one single strategy (or combination of strategies) that will be best, is not an easy task. A lot of experimenting may be required before the best approach emerges.

9
Step 4:
Determining
Training Needs

Introduction

The U.S. Air Force learned a valuable lesson while implementing its TQM education and training programs. They trained the masses and did not achieve significant results. Training and education for their own sake do not pay off. However, education and training united with implementation make for a dynamic combination.

Step 4: *Determining Training Needs*, focuses on the needs of the purchasing department and the supplier community. Although there is a difference between education and training, the balance of this section will not separate the two, but treat them as a complementary package:

Education enables users to understand and answer "why" questions. (Why are supplier partnerships important?)

Training focuses on providing answers to "how to" questions. (How do we go about establishing supplier partnerships?)

Education and training is not a quick and simple process. For example, one major corporation spent six months educating its buyers on developing partnerships. When asked how the process went, an executive indicated that they had had problems. In fact, he said, 90 percent of the original buyers were no longer with the firm—they could not adjust!

The Education and Training Focus

Depending on the purchasing organization, training may involve such basics as writing skills and telephone techniques, or be at a level of sophistication that involves developing supplier relationships and use of statistical process tools. Likewise, the training and education needs of suppliers will vary, depending on their capabilities and their current relationships with the purchasing organization.

Much of the training will depend on the strategies selected. For example, if involving suppliers in the design process is a selected strategy, then training and education for the purchasing organization, purchasing's sister departments (manufacturing, engineering, and quality), and the suppliers in implementing this strategy will be appropriate.

In general, there are two focuses for the education and training effort:

1. Education and training in TQM philosophies and principles. Subjects include quality awareness and the TQM process; quantitative measures to analyze processes; group development skills such as team building; and understanding the change process.
2. Education and training in the implementation of strategies selected. Subjects include the development of strategies (developing partnerships, the certification process); and the development of measurements to monitor improvement in these strategies.

The steps to determine training needs are shown in Figure 9-1.

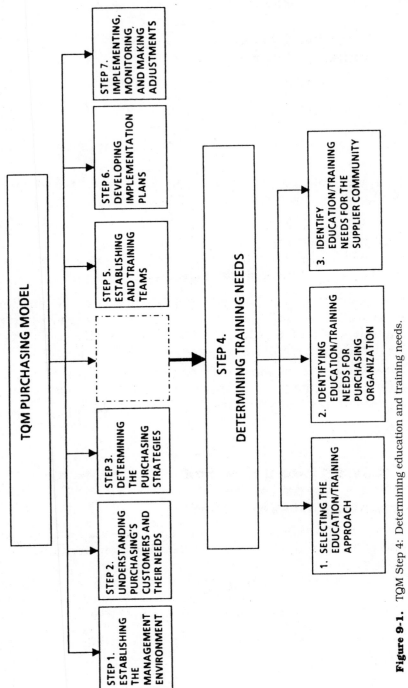

Figure 9-1. TQM Step 4: Determining education and training needs.

135

Selecting the Education and Training Approach

There are several training and education options:

1. Start from scratch and develop your own training and education approach.
2. Hire a consultant to develop and do all the training and education.
3. Buy an existing package and modify it to suit your needs, with your own people doing the training.

One popular method used by those who opt for the third approach is "cascading" education and training, whereby each management level educates and trains his or her subordinates. The advantages of this approach are:

Line accountability. The manager is responsible for making TQM work. Education and training is one of the most important parts of that, so each manager is personally accountable for the education and training of his or her own people.

When you teach, you learn. If you don't understand TQM, it's difficult to explain it to your people. This approach guarantees a group of managers who truly understand TQM.

Identifying Education and Training Needs in the Purchasing Organization

Buyers tend to experience an immense personal and professional change in outlook as they develop long-term relationships with suppliers. They need to learn new concepts to facilitate this process. The Ernst and Young Quality Consulting Group, in their book, *Total Quality: An Executive Guide for the 1990s,* list the following learning targets for buyers:

- Learn to manage supplier relationships with a win-win approach.

- Monitor progress and assist in removing any barriers to an effective working relationship.
- Work with others to evaluate suppliers based on capability and willingness to improve.
- Focus on improving supplier quality, delivery performance, and total cost.
- Take the lead in establishing improvement goals.

Quality Training Methods for Purchasing

The American Productivity and Quality Center of Houston, Texas, recently completed a study on the training methods used for training organizations in quality. Since a large portion of the respondents were purchasing managers, the data collected applies to training purchasing personnel in TQM methods. Over 50 percent of the respondents indicated that they had spent more than one week in quality training during 1989. Almost 70 percent stated that they had attended at least one training session devoted to quality and productivity during the same period.

The most popular general topic for quality improvement is employee involvement. The most popular training topics from the survey in popularity order were:

Employee involvement

Quality management

Productivity management

Productivity measurement

Statistical process control

Rewards systems for improvements

The survey also indicated that most managers insist that those taking the quality training take responsibility for implementing suggested techniques.

The formats for the training favored more traditional methods than interactive teaching or visits to manufacturing plants. The methods that had the greatest effect, in order of impact, were:

Small-group seminars

Lectures

Videotapes

Workbook exercises

Large national conferences

Plant/company tours

Interactive videos

Computer programs

The study also obtained data that ranked the areas that also needed training to assist in improvement:

Strategic quality planning

Productivity/quality measurement

Team building

Customer satisfaction

Reward systems

Service sector productivity/quality improvement

Interpersonal communication

A great resource for the training process is suppliers. Many suppliers are farther along in the process than the customer's purchasing department.

Identifying Education and Training Needs in the Supplier Community

Unfortunately, few suppliers have all the capabilities needed to make them ideal. Ernst and Young's book, *Total Quality,* indicates the following ways in which assistance can be provided to suppliers:

- Recommend sources for education, training, and technical assistance.
- Sponsor or conduct in-house education and training sessions.

- Provide hands-on technical assistance at the supplier's location(s).
- Demonstrate the application of various tools and techniques at purchasing's location(s).

A joint effort with the supplier can greatly enhance the identification of education and training needs.

Summary

One of the constants in TQM is the need for education and training. Regardless of the strategies selected, education and training will be needed, both by the purchasing organization and by the supplier community.

10

Step 5: Establishing and Training Teams

Introduction

The use of teams to achieve continuous improvement is one of the basic principles of TQM. Teams utilize employees at all levels. All the employees need to bring with them is expertise in their field and an ability to work with others. Within a well-functioning team, one person's weakness may be another's strength. Everyone on a team can contribute ideas, plans, and suggestions for improvement and expect those ideas to be heard and then fairly and openly discussed and evaluated. Mature teams are capable of making much better decisions than all but the most brilliant individual contributors. Team efforts produce many tangible as well as intangible benefits. Among the intangible benefits are a sense of belonging and a heightened sense of owning the process.

Teams work especially well in the purchasing world because that environment has become so complex that no individual has the necessary breadth of knowledge and experience to deal effectively with the full range of issues involved in most important decisions.

Improvement Teams Do Not Equal Quality Circles

An early attempt to utilize improvement teams in the U.S. was the concept of "quality circles." They were touted as the Japanese answer to employee involvement, and in Japan, they work. The success or lack of success of quality circles in U.S. industry continues to be debated.

One of the traditional approaches to quality circles was to have a team make recommendations. These suggestions, more often than not, fell into a "black hole," never to be heard about again. TQM improvement teams, however, are different from quality circles. Improvement teams are empowered to implement what they recommend. They are given the resources (people, machines, and money) to ensure that their ideas are successfully used.

Figure 10-1 illustrates several of the activities involved in establishing and training teams.

Team Types and Structures

The type and structure of teams should be flexible. Some organizations have formalized methods for establishing teams, while others form teams on an ad hoc basis. There are no right or wrong types and/or structures.

Under TQM, the purchasing professional will participate in at least two types of teams:

1. *Source-selection or commodity teams.* Usually the purchasing professional serves as the team leader in this type of cross-functional team. The team's objectives are to evaluate, select suppliers, form buying strategies, and implement new supplier/buyer relationships.

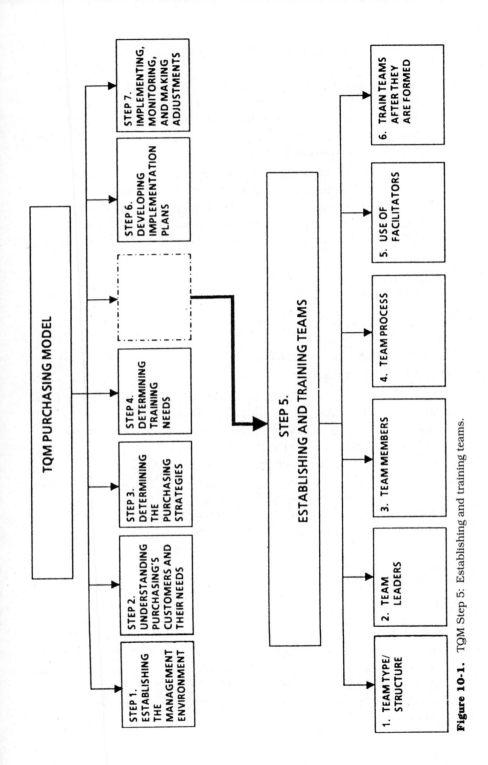

Figure 10-1. TQM Step 5: Establishing and training teams.

2. *Efficiency improvement teams.*   In this type of cross-functional team, the buyer is either the leader or an especially active member. The focus of these teams is either internal (on issues within the purchasing professionals' own company) or external (on issues involving suppliers).

   ▪ Internally oriented teams focus on improving company processes and methods that have a critical impact on the purchasing process. These can be acquisitions processes (requests for proposals), administrative processes (personnel assignments), or miscellaneous processes (requisitioning supplies).

   ▪ Externally oriented teams focus on improving the suppliers' capabilities.

Source-Selection and Commodity Teams

Source-selection and commodity teams are cross-functional, with members from at least the following departments: accounting, engineering, manufacturing, quality and production planning/control. The buyer is usually the leader, but oftentimes the leadership is rotated. Team members have specific roles in achieving the objectives of evaluating, selecting, forming strategy, and/or implementing new supplier/buyer relationships.

An example of the use of these teams is in the production readiness review of a current supplier used by ITT Defense. The members of the team would consist of the buyer and representatives from manufacturing, quality, and production planning and control. The roles of each member are described below.

The Buyer

As the team leader, the buyer has ultimate responsibility for the successful completion of all activities in the process. He or she spearheads the development of plans in support of the specific objective. In addition, the buyer is responsible for evaluating the following:

Documentation to identify, procure, schedule, and control materials to manufacture the product

Identification of single, sole, dual, and alternative sources

Programs to issue and track purchase-order performance by subcontractors

Systems to coordinate and control production schedules and procurement plans for materials, parts, and processes supplied by subcontractors

Coordination between quality and purchasing to assure compliance to specifications and control of material quality

Methods and results reflecting communications between manufacturing and purchasing to identify and place off-loaded and/or subcontracted manufacturing support

Status of any single- or sole-source supplier's labor contracts, including union contract expiration date, strike history, and alternative source development, and rate atonement lead time

How risks will be identified and what actions will be taken

The Manufacturing Member

This member does a detailed evaluation of facilities and equipment planned for production and identifies and evaluates suppliers' control of high-risk manufacturing methods. Further, he or she evaluates:

Manufacturing methods and tooling concepts with the manufacturing methods and facilities

The suitability of documentation to control the fabrication process and changes to the process

Producibility plans to identify and correct manufacturing problems or to optimize methods and reduce cost

Systems to identify and correct manufacturing problems

Test-equipment needed, utilization plans, time phasing of availability, proofing plans, capability, and capacity to meet rate

How risks will be identified and what actions will be taken

The Quality Member

The quality member evaluates:

The organizational structure of the quality assurance group and its compliance with quality requirements

Communication between supplier and customer regarding the quality of the product and statistical quality control techniques

Maintenance of accurate and complete records of inspections and tests performed

Written procedures describing the quality assurance system and quality assurance plan

Procedures for the engineering and manufacturing evaluation and control of designated critical parts

Communication with inspection personnel to assure compliance with quality control policies and techniques

Procedures to evaluate/control deficiencies discovered during inspection of the product

Plans and methods for the first inspection of the first unit to assure compliance with requirements

Systems to monitor processing operations and enforce applicable process requirements

Plans for receiving and in-process inspections

Systems to control and monitor changes to hardware, software, and technical data

Procedures for control of supplier's suppliers, including plans for source inspection and incoming inspection

How risks will be identified and what actions will be taken

The Production Planning and Control Member

This member evaluates:

Procedures used to schedule and load facility

Systems that provide adequate inventory control of all materials and products

Systems to maintain records of labor performance, material usage variance, scrap and rework costs, equipment utilization and analysis of variance to predetermined targets

Processes for selection, training, and assignment of skilled manpower

Reconciliation of actual experience to plan

Processes for making or buying decisions

Systems to coordinate internal production schedules to higher-tier schedules

How risks will be identified and what actions will be taken

Efficiency Improvement Teams

Team Members

Many teams consist of management and hourly employees. Supervisors are often not welcomed as members by other members because the areas targeted for improvement are their supervisors' areas of responsibility and make them uncomfortable. Also, team members are often reluctant to be open in front of their supervisors.

Team Leaders

The team leader is usually appointed but can be elected by the team members. This duty is often rotated. The team leader functions as the leader in team meetings and the implementation of action plans. When appropriate, the leader interfaces with the management on team issues and concerns. He or she finds volunteers and/or makes assignments to individuals or subcommittees as needed.

Team Process

A structured methodology for the team is critical. The standard rules of meeting processes apply and are complemented by a specific process that all teams should use. For example, each team should document (write down on a piece of paper for all to have) the purpose of the team. The purpose is a clear, concise objective and statement of work. In addition, meeting rules should be established, minutes taken and distributed promptly, set meeting times established (usually once a week for no more than one to two hours). Formalized team processes range from six steps at ITT Defense to thirty-four steps at Textron-Lycoming.

Team Facilitators

Facilitators can be used to instill enthusiasm and a sense of mission about improving processes. Why then don't they always succeed? The problem is that typically they possess strong motivation but little control or power. In the December 1990 *Journal for Quality and Participation,* Joanne Sullivan and Tom Werner listed ten "power tools" for facilitators:

1. *Access.* Always consult with the leader of your organization.

2. *Self-management and organization.* Think about strategy and results, not just activities.

3. *Feedback.* Let people know what you are doing.

4. *Tracking and reporting.* Chart results; separate the implementation progress (what the participants are doing) from your own activities (what you are doing).

5. *Consult by walking around.* Get out and about.

6. *Communication skills.* Practice, practice, practice makes perfect.

7. *Simplistic thinking.* Use Tom Peters and Bon Waterman's "smart-dumb" rule: don't be too smart for your own good—it makes you dumb.

8. *Modeling.* People are watching you all the time; be a model and set an example.

9. *Self-care.* Although you care deeply about the people in your organization, you will grow weary of them; have someone else to talk to.

10. *Guts.* State simply what you believe is happening, what you think and feel, and what you would like to see happen.

Training Teams After They Are Formed

We have been successful in providing focused training after groups or teams are formed. Training teams is not easy and requires expertise not usually available to most organizations. Spend the money and obtain the required training even if outside services are needed.

Lessons Learned

The first teams we set up in purchasing taught us valuable lessons:

- Focus the team on a very specific and well-defined objective. The road to that objective should be marked by milestones that will either have a clear accomplishment or completion in a three- to six-month time frame.
- Require a high level of participation by all team members. One approach is to use "coaches" or advisors. A higher-level manager can be assigned to a team with the objective of assisting only, rather than full participation. The advisor will not necessarily attend all the meetings, but will assure that progress is being made and provide guidance as needed.
- Define the measurements for improvement at the outset. Drive for quantitative measurements when possible. This will encourage the team to focus on the basic changes that will achieve improvement.
- At a minimum, train the team leader. The leader in many cases will determine the success of the team.
- Weed out the obvious nonteam members. A nonteam member has sharp things sticking out of his or her sides. Whenever he rubs up against a real team member there will be a lot of blood.

Summary

The use of teams is a basic principle of TQM. For purchasing, the teams need to include other functional representatives as well as suppliers. A final word of caution: The use of the accolade "team of the year" is not appropriate. It can backfire and produce negative reactions. All teams should be judged on their own merits, not one against the other.

11
Step 6: Developing Implementation Plans

Introduction

With the management environment established (Step 1), the needs of the customers identified (Step 2), the strategies selected (Step 3), the training needs determined (Step 4), and the teams in place (Step 5), the next step is to develop implementation plans. (See Figure 11-1.)

There are as many approaches to developing implementation plans as there are strategies. Every company has its own methods of accomplishing this objective. The key to continuous improvement in implementation is summarized in this quotation from one of TQM's gurus, Joe M. Juran.

All quality improvement takes place project by project and in no other way. The critical word here is *project*. We define a project as a problem scheduled for solution—a specific mission to be carried out....The [project selection] process is an essential

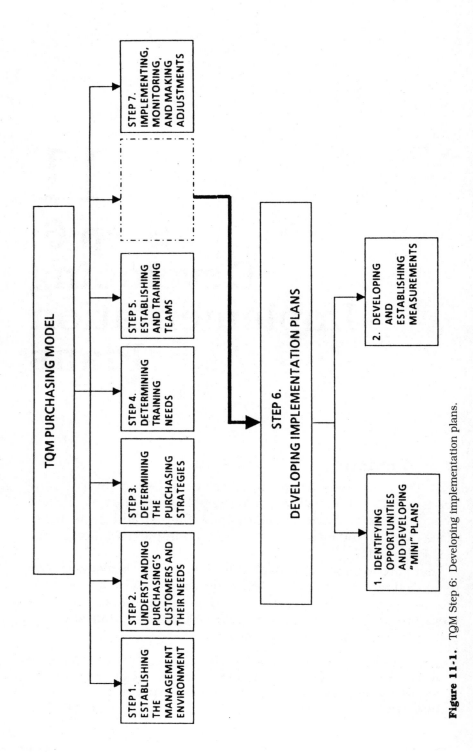

TQM PURCHASING MODEL

STEP 1.
ESTABLISHING
THE
MANAGEMENT
ENVIRONMENT

STEP 2.
UNDERSTANDING
PURCHASING'S
CUSTOMERS AND
THEIR NEEDS

STEP 3.
DETERMINING
THE
PURCHASING
STRATEGIES

STEP 4.
DETERMINING
TRAINING
NEEDS

STEP 5.
ESTABLISHING
AND TRAINING
TEAMS

STEP 7.
IMPLEMENTING,
MONITORING,
AND MAKING
ADJUSTMENTS

STEP 6.
DEVELOPING IMPLEMENTATION PLANS

1. IDENTIFYING
OPPORTUNITIES
AND DEVELOPING
"MINI" PLANS

2. DEVELOPING
AND
ESTABLISHING
MEASUREMENTS

Figure 11-1. TQM Step 6: Developing implementation plans.

part of the company's structured approach to quality improvement.

TQM Implementation Gurus' Views on Suppliers

Having never read a book on TQM without encountering a discussion of TQM's gurus, I feel obligated at least to mention them. This select group of quality experts has been advising industry for years on how it should manage and implement quality. It may be useful to consider their individual approaches, their similarities and differences and, in particular, their views on purchasing, rating suppliers, and single-source supply.

One thing they all have in common is the belief that there are no shortcuts, no quick fixes. Another is their conviction that improvement requires full commitment and support from the top, extensive training and participation of all employees.

Phil Crosby

Crosby is known for his concept of "zero defects." He has written several books, perhaps the most famous being *Quality Is Free* (McGraw Hill, 1979), which sold over a million copies. Crosby's definition of quality is conformance to requirements, and he believes in one standard of performance: zero defects. There is no place in his philosophy for statistically accepted levels of quality, because he believes that this leads to the idea that errors are inevitable and planned for. He offers management a fourteen-step plan to achieve quality improvement.

Crosby views suppliers as an extension of the business and sees most faults as due to purchasers themselves. Crosby believes in rating suppliers and buyers, but holds that quality audits are useless.

Bill Conway

Conway is the least known of the American gurus and is a relative newcomer. After several years at the U.S. Naval Academy, he started an industrial career that led to his becoming president and chairman of Nashua Corporation. He is described as a Deming disciple. Conway does not have a specific definition of quality, but incorporates it into his broad description of quality management: development, manufacture, administration, and distribution of consistent low-cost products and services that customers want and/or need. He focuses on eliminating waste and concludes that the bottleneck is at the top of the bottle—i.e., he believes that the biggest roadblock to TQM implementation is top management. Conway defines six tools for quality improvement.

Conway's call for improvement includes suppliers. He believes that suppliers should be rated, using statistical surveys.

Dr. W. E. Deming

Dr. Deming is considered the father of quality by many in the field. A statistician, he gained fame by helping Japanese companies to improve quality after World War II. Japan recognized his efforts with the award of the Deming Prize to its top-quality companies. Deming's basic philosophy is that quality and productivity increase as variability decreases and that, because all things vary, statistical methods of quality control must be used. He advocates the use of statistics to measure all areas. He is an advocate of employee participation in decision making and believes that management is responsible for 94 percent of all quality problems. Deming has laid down fourteen points for management to abide by, which can be used both internally and for qualifying suppliers.

Deming professes that the inspection phase is too late in the process to discover faults. Inspection allows defects to enter the system through acceptable quality levels. Deming does not hold with supplier rating and is critical of most rating systems. He believes in a single source of supply.

Joe Juran

Juran, along with Deming, is credited with some of the success of Japanese companies. Juran first coined the term *fitness for use or purpose.* He points out that a dangerous product could meet all the specifications, and still not be fit for use. Juran was the first to deal with the broader management issues of quality, an approach that distinguishes him from those who espouse specific techniques. In his view, less than 20 percent of quality problems are due to workers, with the remainder caused by management (compare to 94 percent Deming believes are caused by management). Juran offers a ten-step plan for achieving quality improvement.

Juran believes in rating suppliers, but also advocates helping suppliers improve. Juran does not believe in relying on single sources of supply because he feels that the approach leads to a dulling of competitive edge.

One Tested Approach to Implementation Planning

Most companies have their own standard approaches to the development of implementation plans. The significant action at this stage is to ensure that the plans do in fact link the strategies with the needs of the customer.

An approach I have had some success with includes developing and implementing plans that will yield short-term (three-to-six-month) results in tandem with plans that are more long term (one year or more) in execution and results.

We called the short-term plans "quick-wins." They were sized for incremental, step-by-step approach, low initial risk, low cost, high return, consistency with long-term plans, and realistic doability. The feedback from these plans was intended to validate the long-term plan and the overall, macro-plan—i.e., was the macro-plan reachable, and what went wrong or what went right with the short-term plan?

I had an opportunity to use this method in both my purchasing and operation strategy experiences. Although it was a sound approach, it was hard work. Without a steady champion of the process (preferably the CEO) it possibly would not survive in other corporate settings.

Identifying Improvement Opportunities and Developing Miniplans

The first team action is to identify improvement activities for each strategy selected. A set of tasks are then identified for each activity. These tasks focus either on a single issue or on a set of related issues. All tasks should be structured to be manageable and self-contained. The deliverable result expected from each team is a "mini-implementation plan." The miniplan contains the following:

- A flow chart of the process under review to include the process boundaries, outputs, customers, inputs, and suppliers
- Identification of the process owner and process members and the role of the process members
- The alternatives considered
- A selection of the most probable alternatives with the most probable attributes (time frames, people resources, and money) scoped out
- The worst case (most expensive) scenario
- Tangible savings expected

These plans are then presented to a top-level steering committee for evaluation. The plans are designed to allow the steering committee to determine the feasibility of implementation and validate ability to meet the desired goal.

Responsibility for developing these plans is assigned to the team in addition to their regular responsibilities. While this may create some difficulty, it is necessary in order to utilize the best people from the company (who are usually the same people who run the business).

Applying this approach to the purchasing strategies will require the use of many of the key purchasing personnel, as well as the use of other key functional employees. One method to spread the effort is to ask selected suppliers to participate at this stage.

Developing and Establishing Measurements

The objective of the development of measurements is to determine those measurements needed to understand and improve the process. One common approach is to determine where you are in any particular area and then set an arbitrary improvement percentage. In many situations the selection of a target does reflect an attainable goal, but may not always consider special situations that can require a step-by-step approach to improvement.

In setting many of our goals, we learned to include the particular conditions that existed. For example, in measuring the effectiveness of our purchasing group, we had to consider the type of buying for the period being measured. If we were buying parts for development purposes, improvement goals had to reflect that type of buying. As we moved into production buying, the goals had to be adjusted to indicate improvement for this category. The lesson we learned is to not set arbitrary improvement targets, but to match the targets to the real world.

A measurement method developed by Frank Colantuono, director of planning at ITT Defense, requires that the output of each process be measured according to three basic elements: productivity, quality, and timeliness. All measurement data must be readily available, easily compiled, retained weekly in a graphic format, and recognized as a fundamental part of the system. This measurement process is not for the faint of heart, as it exposes current levels of performance to the organization. See Appendix B for a detailed explanation of this and other measurement approaches.

12

Step 7: Implementing, Monitoring, and Making Adjustments

A "program of the month" tends to have limited survival, and if left unattended will become ineffective or fade away. It is necessary to perpetuate the continuous improvement process forever. This constant reinforcement supports the idea that TQM is not a program but a new day-to-day behavior for all.

The final step in the model is to implement the actions identified. Monitoring the results and making adjustments are a natural outcome of implementation. (See Figure 12-1.)

Creating a formal organization can be a plus in perpetuating the improvement process. One method that will enhance reinforcement is the use of an advisory group or steering committee. An application of this concept includes the assignment of one member of the group to serve as a "coach" to each team. This provides a stimulus for the leader who periodically meets with the coach to discuss progress. It also provides the leader with support and backup as he/she periodically updates the advisory group.

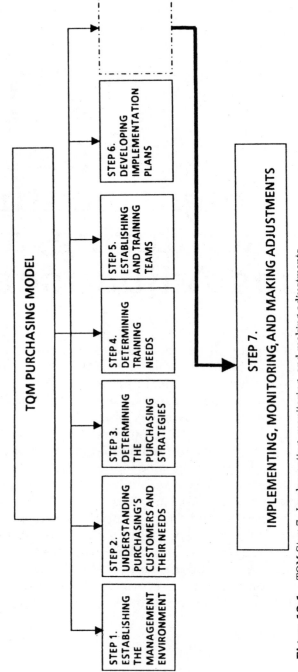

Figure 12-1. TQM Step 7: Implementing, monitoring, and making adjustments.

Appendixes

Appendix A
Does TQM Apply to Defense Contractors?

Introduction

Ansari and Modarress, in their book, *Just-In-Time Purchasing*, make the following statement regarding the application of JIT to defense contractors: "Since the U.S. government is unlikely to change its policies, the only solution is to give up JIT [TQM] practices or government contracts."

Although this view can be adequately substantiated, I have an alternative position. I had the opportunity of spending several years in a major defense company, with my most recent assignment being that of purchasing manager. Based on my experience, it is my opinion that all the principles of TQM *do* apply to defense contractors.

DoD Actually Pushes TQM

Current attitudes toward TQM in the Department of Defense are so favorable that its application there may even be more appropriate than it is in some commercial companies. As an example, in 1989 a joint U.S. Air Force and industry group Process Action Team (PAT) published their findings and recommendations on TQM impediments. The PAT identified twenty-six impediments and made recommendations for each that included changes to laws relating to government acquisition and associated policies—regulations and practices that impede the Department of Defense and related industries from effectively pursuing the type of relationship beneficial to TQM goals. Several recommendations dealt directly with the customer-supplier relationship.

Impediment:	The Competition in Contracting Act (CICA) of 1984 requires all procurement actions to be awarded on the basis of full and open competition. This results in significant non-value-added costs as a result of a proliferation of bids and multiple suppliers. The act also makes it easy for companies to file protests over contract awards. This policy is in direct contradiction to TQM principles.
Recommendation:	Form a PAT to recommend changes to CICA.
Impediment:	Government and industry's emphasis on meeting schedules rather than on achieving quality leads to contractors responding to RFPs with unachievable schedules. Contractors are fearful of appearing unable to meet the government's requirements. Poor quality, delinquencies, and cost overruns result.
Recommendation:	DoD should emphasize the use of more realistic schedules and provide training on the relationship of quality to schedule.
Impediment:	Mixed signals to industry from government, with each military service practicing TQM in its own way.
Recommendation:	Establish a formal forum for TQM interchange between DoD and industry.
Impediment:	DoD cost-based pricing policy fosters inadequate incentives for sole-source contractors to improve quality and reduce costs. The policy uses past per-

formance for allowable costs and establishment of overhead rates.

Recommendation: Establish a PAT to improve the process.

In addition, contractors continue to be asked by a variety of government agencies to identify regulations and government practices that ask them to oversee, but not to add value, to the contractor's activities.

Defense Contractors' Material and Purchasing Personnel Speak Out

In recent symposiums (1990 and 1991) sponsored by Coopers & Lybrand, I had an opportunity to meet with more than eighty material and purchasing professionals from the defense contracting world.

Coopers & Lybrand concluded that the CICA was written for the purpose of awarding DoD business based on competition (and the more competition the better). The defense industry experiences the negative effects of this short-term policy. The end result is that partnerships within the industry are not happening. It was this situation that prompted Coopers & Lybrand to begin the process of "becoming a part of the solution" by sponsoring the symposiums.

There was strong consensus among the group that prime contractors are forced to deal with suppliers as adversaries, sacrificing quality, delaying schedules, and increasing costs to the government. However, there was also a belief among many (including this author) that TQM principles *do* apply to the defense world. Nor is the application of these principles for continuous improvement in direct contradiction to the principle of ensuring that all suppliers be given a fair shake at bidding. In fact, the concept of reducing the supplier base is actually very appropriate, and, done properly, will continue to ensure that all suppliers have an equal chance to participate.

One outcome of the seminar was the identification of improvements/issues that needed to be addressed and resolved:

1. *Converting adversarial relationships into working partnerships.* Congress, DoD, and contractors should stop thinking of their suppliers as adversaries. Contractors would like to see their customers initiate working partnership relationships with them, based on trust and respect.

2. *Reevaluating the use of competition and reassessing its effectiveness.* There is some doubt as to whether competition, as it is legislated today, has added much value to the acquisition process. The CICA of 1984 represents a congressional short-term fix to a perceived, poor acquisition process. The DoD Competitive Advocacy Program is no longer effective, and the defense industry is feeling its shortcomings. The low-bidder approach does not necessarily result in the lowest overall cost to the DoD. A congressional committee is needed to work with organizations such as AIAA to develop a new acquisition approach.

3. *Making quality a priority (other issues will fall into place).* With quality the focus, there will be reduced cycle time, rework, returns, warranty costs, and lead times. Quality builds customer loyalty and is key to our national competitiveness.

4. *Involving suppliers up front at the design stage and throughout the process.* Teams need to be developed that include contractor functions and supplier functions. The old "total-control" environment, maintained by requiring at least half-a-dozen signatures to accomplish anything, is being replaced by a total-quality focus on satisfying customer needs by using quality improvement concepts and employee empowerment.

5. *Identifying the non-value-added processes and removing them.* The acquisition process is controlled by the politicians. It thus provides a great opportunity for finding non-value-added regulations, paperwork, rubber-stamp and sign-off requirements. If all the DARs and FARs were followed to the letter, would DoD and Congress get what they wanted? Probably not.

6. *Knowing who the quality suppliers are and rewarding them with more business.* When a contractor has a good supplier and that supplier makes improvements to its process, the contractor gets better products. But the contractor cannot encourage its suppliers to continue this by awarding them more business.

Instead, the acquisition process pushes the contractor to re-compete each buy and start the process over again with another supplier. Under this business relationship, how can a contractor expect its suppliers to want to work with them to make improvements in the future?

7. *Converting the acquisition system from a short-term to a long-term orientation.* Acquisition needs to learn to make decisions based on the long-term effects and benefits. Create a business environment that facilitates long-term thinking. Assure long-term business arrangements to suppliers based on their total-quality performance.

8. *Fostering supplier partnerships that lead to good communications and mutual trust between customer and supplier.* Most communication in this industry is in a downward direction. DoD and Congress need to create an atmosphere of mutual trust. Examples of poor communication can be seen in many Statements of Work (SOW) that necessitate *re*work because they are so unclear. The auditing process could also be a value-added TQM method, but as it stands today, is a lose-lose situation. It doesn't use information to improve the process, but instead leads people to hide mistakes and errors out of fear of retribution.

Perceived Roadblocks for Defense Contractors

The following roadblocks, preventing the application of these principles, are the most frequently cited by defense contractors:

Roadblock: TQM principles are in direct contradiction to the Competition in Contracting Act of 1984. The act was written to promote the award of Department of Defense (DoD) business on the basis of competition (and the more competition the better).

Approach: Stop focusing on what cannot be done and concentrate instead on what can be done. It is time for action.

Roadblock: Defense contractors are prevented from awarding contracts to preferred suppliers.

Approach:	Implement a certification process open to all, and stipulate that only those who are certified will be allowed certain business. (Note: The army has a contractors' performance certification program that is by invitation only and is voluntary. Certified contractors will be favored in future awards.)
Roadblock:	Defense contractors must provide for minority suppliers.
Approach:	Set aside specific dollar amounts for minority suppliers and award those dollars only to them. Don't include this money (generally minor) in the TQM process.
Roadblock:	Defense contractors must allow all potential suppliers to bid.
Approach:	Allow all to bid for the initial award, then down-select suppliers, using well-published ground rules to reduce the supplier base.
Roadblock:	Defense contractors have difficulty developing long-term relationships because suppliers will only be around until the contract is completed.
Approach:	Most suppliers are willing to work three to five years for a reasonable profit in a program that requires their unique technology.
Roadblock:	Suppliers are reluctant to get involved at the design stage because there is no commitment that they will be awarded the business.
Approach:	Make an up-front commitment that suppliers will be awarded the business if the contractor gets the business. Down-select suppliers based on technology before design effort is required.

Summary

TQM concepts and approaches do apply to this industry and DoD is emphasizing its use. The focus must be on what has to be done to use these principles, rather than on developing more roadblocks.

Appendix B
Measurement Techniques

Introduction

"If it doesn't get measured, it doesn't get done," says General Loh of the U.S. Air Force. With this concept as a theme, this appendix outlines alternative measurement techniques for the purchasing function. The focus is on new ways to measure purchasing performance in light of TQM. The traditional measurements (on-time delivery, cost savings, etc.) continue to be valid and are complemented with these new concepts.

Use Metrics Cautiously

The use of metrics to measure performance can be viewed as threatening to the supplier or individual being measured. Unless the measurements are clearly focused on fostering improvement (rather than on crucifying the person or supplier being measured), the effects can be negative indeed.

(In setting up metrics for our operation, my boss, a vice president, asked me to indicate tactfully that he was concerned about exposing to his peers and his boss all the problems that existed in the operation. To accomplish this I generated a cartoon illustrating how he was opening himself up for all to see and critique.)

To avoid the devastating potential of this type of measurement, several actions can be taken. If, for example, a measurement will be of a purchasing supervisor, ask the supervisor to participate in the process, and have him or her identify those areas he or she believes are critical. The measurement will probably be something that the supervisor can control. If he or she picks an area you do not believe critical, the supervisor may be focusing on the wrong area. Allow the selection process to require an agreement between the person being measured and the person receiving the measurement information. During presentations by the supervisor, ensure that he or she is not punished for a process that is not improving. Provide an opportunity to discuss the area and focus on ways to improve the process.

TQM-Type Measurements

Under TQM, several "new" individual measurements are used to complement the traditional ones. Examples of these TQM measurements include:

- Percentage of business not inspected
- Percentage of business with certified suppliers
- Supplier's responsiveness to problems
- Percentage of purchasing budget spent on supplier training

Benchmarking

Benchmarking is a strategy that many companies use to improve performance (see Chapter 8). It is also a "new" method for measuring performance. Benchmarking metrics use many of the traditional indicators (on-time deliveries, etc.) and many of the new indicators (number of problems solved by suppliers, number of suppliers involved in design, etc.).

According to Litton Guidance and Control Systems, every benchmark is compared at some regular interval against a recognized world leader, to determine relative position. The owners of the process or subprocess being benchmarked must collect their own measurements of health and vitality to make the comparison.

To determine the items to be benchmarked, the exact process must be flow-charted, step by step. Then specific points should be selected in the process or subprocess that display its true health and vitality. Any benchmarking measurement not made for the purpose of decision making (or to satisfy a customer) must be viewed as wasteful.

Benchmarking can be done as frequently as past history dictates in order to establish a statistically meaningful data base, or as frequently as the industry average or as the world-class leader.

The selection of the candidates to be benchmarked will often come from your gut feelings: trust them. Other methods include the use of personal contacts in other companies, data from trade journals, past Baldrige winners, interviews with your customers, and other creative areas.

If your first attempt at benchmarking shows you are better than the company being benchmarked, reexamine your decision-making process and try again. If you determine that you *are* world-class, let the world know. Share the data with everyone in the process or subprocess.

Litton's Ten-Step Benchmarking Process

Planning:	1. Identify what is to be benchmarked.
	2. Identify comparative companies.
	3. Determine the data collection method to be used and collect the data.
Analysis:	4. Determine current performance gap.
	5. Project future performance levels.
Integration:	6. Communicate benchmark.
	7. Establish functional goals.
	8. Develop action plans.
Action:	9. Implement specific actions and monitor progress.
	10. Recalibrate benchmarks.

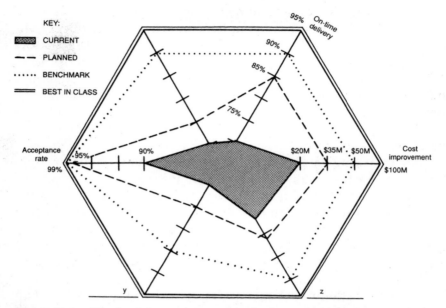

Figure B-1. "Arachnids": metrics reveal performance. (*Hamilton Standard*)

Hamilton Standard's Use of "Arachnids" to Show Its Performance

Figure B-1 demonstrates a unique way to illustrate performance in relation to benchmarking. Hamilton Standard's use of "arachnids" provides a simple yet comprehensive method of showing its own performance as compared to its benchmark and the best-in-class.

The Comprehensive Index Approach

Indices are often used in developing a supplier rating system. An elaborate and complete one is outlined in Frank Caplan's book, *The Quality System*. This approach provides separate indices for each key factor, with an overall index that combines them using a 1.0000 base.

The indices for this plan are Quality, Delivery, and Service—with a composite total of these three, properly weighted. The weighting factors must be internally generated and usually involve purchasing, engineering, manufacturing, and quality. A typical weighting factor has Quality = .55; Delivery = .25; and Service = .20. When Cost is another index, the weighting is Quality = .40; Cost = .35; Delivery = .15; and Service = .10.

The system includes delivery factor demerits for early and late delivery and includes provisions for Just-In-Time efforts. The service factor is the most variable in content and has a merit and demerit rating for parameters such as cooperation in pre-award surveys.

In this system, supplier ratings are usually published monthly or quarterly, with updates on demand.

Value-Added

Purchasing should be viewed as a value-added function to the product and not one that just adds overhead. This means putting less emphasis on such efficiency measures as the cost of issuing a purchase order and more emphasis on such strategic issues as early supplier involvement in product design and competitive sourcing. Examples of a new set of measurements include:

- Percentage of purchasing's operating budget spent for on-site visits with suppliers
- Percentage of operating budget spent on supplier training
- Percentage of operating budget spent on buyer education and training
- Number of purchasing professionals with technical or engineering backgrounds

The Work Value System to Measure Productivity

"A work value system measures productivity," according to Michael La Placa, director of purchasing at Lockheed, as quoted in a recent *Purchasing World* article by Thomas F. Dillon. The system is based on assignment of work values, in fractions of an hour, for selected purchasing jobs. Efficiency ratings for individual buyers and for the department are a result of comparing work values. Buyers are not rated against each other; they are rated against past performance. The system is never used as a disciplinary tool nor to pressure a buyer. It does let the buyers know how they are doing.

One of the first steps is to build a statistical foundation from data on purchase orders placed, time required to place them, their dollar value, and types (off-the-shelf items, specials, etc.). Using worksheets, buyers keep track of each day's work during a test period. Another analysis focuses on the volume and type of purchasing documents purchased, with a third study showing the different processing requirements for different valued purchase orders.

Based on this information, a work value can be computed for each of the different values and types of purchase order requisitions. These values are validated by comparing the total number of order requisitions processed with the total purchasing payroll hours for one year.

From this information Lockheed knows in theory how long it should take a buyer to process the mix of orders he or she has to handle during the week and can be used to forecast demands and measure improvement of the purchasing group.

Comparable Productivity

Integraph Corporation's corporate director of procurement and transportation, E. Bromley Sweet, developed a process several years ago for measuring and comparing purchasing productivity. His model can be used to compare several divisions' purchasing performance. The model was illustrated in the March 7, 1991, issue of *Purchasing Magazine*.

Sweet feels that the model addresses two important areas needed in performance measurement: (1) measurement of purchasing from the perspective of achieving recognized goals and objectives; and (2) provision for comparing purchasing units separated by geography, type of manufacturing operation, and types of products.

Four factors are used in the comparison:

1. *Preferred procurement.* This factor measures how well purchasing follows a procedure that is recognized as being innately good. Components of this measure include such things as purchases from other corporate divisions, use of corporate and divisional contracts, and use of customer-directed sources.

2. *Supplier performance.* This factor combines a supplier quality rating and delivery rating and divides by two for an average.

3. *Profit contribution.* This factor was given a deliberately lower weight than the other factors. Sweet wanted the model to emphasize supplier performance, quality of purchase transaction, and then profit contribution. The factor includes net savings, cost avoidance, purchased earned discounts, and transportation savings.

4. *Performance factor.* This factor compares a division's volume performance in terms of line items, dollars per purchasing department per person, and dollars per purchasing department hands-on buyer. This gives a comparison of the comparative work volumes handled by each unit.

The model generates a single index and has reports that show performance for each division plus a comparative performance factor.

ITT Defense's Productivity-Quality-Timeliness Approach

Frank R. Colantuono of ITT Defense developed a measurement process that has been under way for several years and has produced significant results. He uses this method as part of TQM in Production and Inventory Control (P&IC). The system allows for the baselining of P&IC's current performance level.

Companies that are striving for continuous improvement often have difficulty in applying the TQM concept beyond the boundaries of personnel covered by engineered standards. The basic problems are the lack of a baseline to measure the impact of improvement ideas once they are implemented, and the lack of a model to be followed in these same areas.

The process may be applied to an entire organization or to any organizational element. It can also be used as a tool to foster continuous improvement. The systematic approach is a hybrid process developed by Mr. Colantuono based on techniques used by the American Productivity Center in Houston, Texas, on articles from the *Harvard Business Review,* and on more than twenty years of progressive management experience in P&IC.

At ITT Defense, the basic task at hand was to assure that:

1. All P&IC customers were being serviced in a way that would allow those customers to continuously improve their process
2. Non-value-added work in P&IC was identified and eliminated
3. The remaining work was done by P&IC in a quality-conscious, efficient, and effective way.

Below is a detailed description by Mr. Colantuono of the steps involved in the measurement process he developed and implemented at ITT Defense:

Step 1: Identifying the Customers

"The identification of our customers was accomplished by analysis of workflows of P&IC activities. These clearly indicated where a product or service was delivered by P&IC to an internal customer. Examples of such activities are the identification of material requirements that will require purchasing action, the delivery of kits of materials to manufacturing, and the processing of vendor materials from the receiving dock to the incoming inspection area. Once our customers were identified, dialogue began to identify the important aspects of the product or service that P&IC delivered to them. Our customers also identified what P&IC was doing or not doing that was having a negative impact on their process. The results became action lists for each P&IC organization to focus on and improve."

Step 2: Fact-Finding Meetings

"This step involved a series of working meetings with the respective P&IC functional supervisors, their managers, and the department director. The process began with the area supervisor outlining the departmental statistics: the number of people in each job category, a complete step-by-step description of how each job is carried out, and a statistical history of the group's measured level of quality, productivity, and adherence to scheduled activities.

Each functional group meeting was carried out in less than two hours. The process of describing how each job is actually done was carried out in great detail and required thorough questioning on the manager's part to uncover the waste in the process. (If you have never gone through this process before, you can be sure that waste is there; the challenge is to discover and eliminate it.) One example found during this process was the needless copying of supplier packing slips. Both the original *and* a copy were sent to the next work center, but the next work center always threw the copy away; it only needed the original. Once that wasteful activity (the copying of the packing slip) was uncovered and eliminated, some value-added task had to be substituted or resources reduced in order to benefit from the discovery. (If no value-added work is put in to replace the waste eliminated, the employee may simply take longer to do the remaining work.)"

Step 3: Gathering Baseline Information

"This step can actually be done earlier in the process but cannot be done any later than this point. Baseline information must be gathered in order to measure the effect of the elimination of waste and of the ideas for improvement on the three basic management parameters. Those parameters are:

1. *The hourly productivity of the functional group.* An example would be the numbers of line items picked by the stockroom kitters divided by the number of hours worked by those kitters (e.g., 2000 line items picked last week divided by 100 hours worked last week = 20 picks per hour).
2. *The quality of output of the group.* This can be measured by

computing the amount of "good" output divided by the total output for the week. An example would be that same stockroom kitting function where 2000 line items were picked. A statistical process control (SPC) work station has been set up after the kitting operation which measures the amount of errors in the process. Thus, by subtracting the number of errors from the total line items picked, accuracy can be determined (e.g., 2000 line items minus 20 errors = 1980 good items divided by 2000 = a total of 99 percent).

Knowledge of the number of errors is only the starting point; the goal is to eliminate errors. The methodology employed by ITT Defense is the Deming Cycle of "plan-do-check," whereby the "check" activity is enhanced through Pareto analysis of the cause of the errors found. Once root causes for errors are found, action lists are used to eliminate the causes.

3. *The timeliness of the delivery of the product or service.* If P&IC does not perform its value-added task on time, our customers' processes are disrupted. The third measurement, therefore, is timeliness, which measures how many things P&IC, by functional area, does as scheduled. My example would again focus on stockroom kitting, where two measures are kept: the number of line items kitted per schedule, and also the number of completed kits that resulted from those line items. P&IC's customer does not work on individual parts but on sets of parts, called kits. (Example: the number of kits required per MRP schedules versus the number provided by stockroom kitting: 95 kits delivered divided by 100 kits scheduled to be delivered per MRP = 95 percent timeliness.) Pareto analysis is also used to understand a missed schedule and the applicable corrective actions. This is the continuing process whereby measurements are done weekly, action meetings are conducted quarterly, and the generation of ideas to improve P&IC activities through the elimination of waste and the elimination of errors in the value-added work is a daily quest.

The process has proven invaluable for P&IC at ITT Defense. Our internal customers were asked to define and assist in developing the measurement criteria for how P&IC services their needs. The functional supervisors and managers in P&IC have gained a more thorough awareness of the impact of their groups' activities on their internal customers. They have also gained quantitative feedback as to the impact of their continuous improvement activities on the productivity of their groups, the quality of the group's output, and the on-time delivery of the product or service."

Appendix C
Purchasing and the Malcolm Baldrige Award

Introduction

In its few years of existence, the Malcolm Baldrige Award has already become a major player in the world of TQM. Many executives believe that it represents an extremely positive step for the U.S. As J. Junkins, CEO of Texas Instruments, puts it: "If you measure yourself against the criteria laid out by the Baldrige Award, you have a blueprint for a better company."

However, the Baldrige Award has also emerged as a topic of debate. Baldrige is not just good news but also bad news, according to Philip B. Crosby.

- The award's credibility came under fire in 1990, when—at the very moment when the U.S. auto industry was generally perceived to be having serious quality problems—the Cadillac Division of General Motors was selected to receive the award. I believe that the Baldrige Award gurus were very cognizant of the apparent contradiction in this situation, and had conducted an extremely thorough site visit to validate the process.

Cadillac didn't help to clarify the situation by using the award in its commercials in a way that suggested the award was for the entire G.M. company.

- Motorola created a major stir in 1989, when it announced that all of its suppliers must compete for the Malcolm Baldrige Award. When its 3600 suppliers learned of the new rule, 200 of them immediately dropped out as Motorola suppliers.

- Purchasing professionals remain very divided on the subject of whether the Baldrige criteria should be used to standardize the supplier certification process.

Regardless of the outcome of the controversy, the Baldrige Award is certainly having an influence on purchasing. If a company decides to apply for the award, the application specifically requires it to address purchasing activities with suppliers. The award can have an impact on relationships with suppliers because the most significant use of the award examination to date involves self-assessment and assessment of suppliers.

The award was designed as a value system, an education-and-communications tool, a vehicle for cooperation, and a device to help evaluate quality standards. Purchasing professionals need to be familiar with this award as the trend toward its use continues to escalate. A review of the award's history follows to shed some light on its gathering importance.

The Baldrige Award Process

The Origin of the Award

The late Malcolm Baldrige was a well-respected secretary of commerce under the Reagan administration and a strong advocate of a national quality award. On August 20, 1987, primarily as a result of Baldrige's efforts, President Reagan signed Public Law 100–107 creating the Malcolm Baldrige National Quality Award, and, in the process, launched a national quality initiative.

The purpose of the award is threefold:

1. To promote quality awareness

2. To recognize quality achievements of U.S. businesses
3. To publicize successful quality strategies

The secretary of commerce and the National Institute of Standards and Technology (NIST) are responsible for developing and administering the awards, with cooperation and financial support from the private sector.

Eligibility for the Award

Only companies incorporated and located in the United States are eligible for the award. Qualifying companies may be privately or publicly owned. Up to two awards may be given in each of three categories:

1. Large manufacturing companies or subsidiaries. Eligible subsidiaries are defined as divisions or business units of larger companies.
2. Large service companies.
3. Smaller companies engaged in manufacturing or services. Small companies are those firms having from 25 to 500 full-time employees.

The Application Process

The 1992 Baldrige Award criteria contain the information needed by companies to apply for the award. Awards are made on the basis of a three-stage process:

Stage 1: The Written Application.   After the applicant is advised of eligibility to compete (based on its submission of an eligibility determination form), the applicant completes an application report consisting of eighty-nine different subject areas.

The application report is then reviewed by a board of examiners:

1. *First-stage review by at least four examiners.*   The examiners, who do not confer at this stage, highlight strengths and weaknesses. A panel of judges then determines which applicants should be referred for consensus review.

2. *Consensus review by at least four examiners.* (Generally these are the same examiners who did the first-stage review.) A panel of judges then determines which applicants should receive site visits.

Stage 2: Site Visits. An on-site verification is conducted by at least five examiners (four of them generally being the same examiners who reviewed the written application). While examining the written application, the examiners have indicated those areas that need site verification, either because the application was not specific enough or because the points made require on-site confirmation.

Stage 3: Selection of Winners. A panel of judges reviews all data and information from the written application and site visit reports and recommends award recipients to the National Institute of Standards and Technology. The institute then makes recommendations to the secretary of commerce, who makes the final decision.

All applicants receive feedback reports.

The Examiners

The board of examiners is comprised of quality experts (including retired quality professionals) selected from industry, professional and trade organizations, and universities. Those selected meet the highest standards of qualification and peer recognition. Examiners take part in a preparation program based on the criteria, scoring system, and examination process.

The Confidentiality of Information

All the examiners sign an agreement promising nondisclosure of information on the applications.

To date, the applications have been secure from the Freedom of Information Act. At least two attempts to get copies of examination results have been unsuccessful, based on rulings from federal judges who refused to release the information.

To protect this process, everything received by the examiners is shredded immediately after the process is completed.

Qualifying for the Award

Criteria are rigorous, with an emphasis on quality achievements, not just systems for quality improvement. The award examination addresses all aspects of a total quality management system, as well as quality improvement results in seven categories of 28 examination items. The 28 items are grouped under sets of subcategories which represent the principal components of the seven categories. The 28 items are further broken down into 89 specific areas that require information in the application. Each category is assigned a number of points. All seven categories total 1000 points, and apply to all applicants. The seven major categories include:

1. *Leadership.*   The senior management's success in creating and sustaining a quality culture (9 percent of total score)
2. *Information and analysis.*   The effectiveness of the company's collection and analysis of information for quality improvement and planning (8 percent of total score)
3. *Strategic quality planning.*   The effectiveness of the integration of quality requirements into the company's business plans (6 percent of total score)
4. *Human resource development and management.*   The success of the company's efforts to utilize the full potential of the work force for quality (15 percent of total score)
5. *Management of process quality.*   The effectiveness of the company's systems for assuring quality control of all operations (14 percent of total score)
6. *Quality and operational results.*   The company's results in quality achievement and quality improvement, demonstrated through quantitative measures (18 percent of total score)
7. *Customer focus and satisfaction.*   The effectiveness of the company's systems to determine customer requirements and demonstrated success in meeting them (30 percent of total score)

The Scoring System

The scoring system is based on three evaluation dimensions:

1. *Approach.* The methodology used by the applicant to achieve the results. It includes the degree to which the approach is prevention-based and the effectiveness of the tools and techniques used.

2. *Deployment.* The extent of application of the approach. It should include all transactions with customers, suppliers, and the public.

3. *Results.* The outcomes and effects. It includes the quality levels reached and the rate and breadth of improvement.

The scoring scale goes from 0 percent to 100 percent, in increments of 10 percent. The scoring extremes and midpoint of the guidelines are as follows:

A 0 percent evaluation. Indicates that there is no system evident (anecdotal) and the approach is not capable of yielding results

A 50 percent evaluation. Indicates that there is a sound approach, deployed sufficiently well to represent meaningful progress, and having some positive results

A 100 percent evaluation. Indicates that the approach is excellent, in full deployment and has achieved exceptional quality levels

(Note that the "anchor" point for a good system is 50 percent.)

Applicants are not advised of their final score, and an interpretation of the total scoring is not readily available. However, using the percentage rating can be a rule of thumb for final evaluation. For example, a 50 percent overall score is more or less equal to a company receiving 500 total points. Using this approach, the 1990 distribution of the written scores was as follows:

Range	Percentage of Applicants in Range
0–125	0
126–250	7
251–400	19
401–600	52
601–750	20
751–875	2
876–1000	0

The scoring trends are moving toward lower scores, with site visits for those that are in the 600+ range.

Baldrige Award Statistics

From its inception in 1988 to 1991, a total of eighteen awards was possible (at the rate of six per year). During this period, however, only nine awards were given. One conclusion that can be drawn from this is that the criteria were too demanding. Another is that U.S. companies are not up to acceptable quality standards. Without getting into this debate, a review of available statistics indicates that the award is beginning to get greater use and exposure.

	1988			1989			1990			1991		
	(1)	(2)	(3)	(1)	(2)	(3)	(1)	(2)	(3)	(1)	(2)	(3)
Manufacturing companies	45	10	2	23	8	2	45	6	2	38	9	2
Service companies	9	2	0	6	2	0	18	3	1	21	5	0
Small businesses	12	1	1	11	0	0	34	3	1	47	6	1
Totals	66	13	3	40	10	2	97	12	4	106	20	3

(1) Applications submitted
(2) Sites visited
(3) Number of winners

Winners:	Manufacturing	Service	Small Business
1988	Motorola Westinghouse (Nuclear Div)	None	Globe Metallurgical
1989	Milliken Xerox (Business Products/ Systems)	None	None
1990	General Motors IBM Rochester	Federal Express	Wallace Company
1991	Selectron Corp. Zytec Corp.	None	Marlow Industries

Manufacturing and service applicants pay a $4000 application fee, plus $1500 for each supplemental section, and submit a maximum of 75 single-sided pages of answers to the 89 question areas. Small businesses pay $1200 and have a 50-page limit.

The effort generated in applying for the award varies. Xerox created a 20-person task force and spent $800,000 to gather the necessary data covering the activities of its 50,000 employees. Motorola devoted 20 man-months to preparing the application and spent $5400 on filing fees and another $7000 to host the Baldrige auditors. Meanwhile, Corning put in 14,000 man-hours and, even though they didn't win, felt the race was worthwhile. In contrast was the effort put out by Globe Metallurgical, which has only two plants, 240 employees, and $155 million in sales. One or two of Globe's four senior executives sat down with an Apple computer and completed the whole application in one long weekend.

The Debate on Baldrige

One of the quality gurus, Philip B. Crosby, has stated that the award is a "good news/bad news idea." The good news is that the award is on quality. The bad news is that it leads to the practice of executives not having any personal responsibility for quality and delegating quality to the quality department.

Crosby's criticism centers on the fact that the award is self-nominating, rather than arising from a nomination process in which customers put forward the names of companies and/or suppliers that deserve recognition. He also complains that the award entails no usable definition of the word *quality*.

Those who disagree with Crosby include Curt Reiman, the director of the Baldrige Award, who indicates that the definition of quality is clear in the award criteria, which explicitly define quality results and closely ties them into customer satisfaction.

Purchasing and Baldrige

If the view is that the supplier is an extension of a company's activities, then all aspects of the award apply to purchasing.

Nevertheless, purchasing activities are specifically addressed in four of the seven categories:

#2.0 Information and Analysis

#3.0 Strategic Quality Planning

#5.0 Management of Process Quality

#6.0 Quality and Operational Results

In the four categories, purchasing has a direct influence on 22.5 percent of the total scoring points (225 out of 1000 points). Figure C-1 lists those areas.

Supplier Quality (Item 5.4) and Supplier Quality Results (Item 6.4) are devoted exclusively to supplier activities. The two items are linked with the results in Item 6.4 by virtue of being derived from quality improvement activities described in Item 5.4.

Item 5.4 Supplier Quality

The 1992 application guidelines ask applicants to "Describe how the quality of materials, components, and services furnished by other business is assured, assessed, and improved." Figure C-2 highlights the areas addressed in this item.

Experts' comments regarding the various strengths and weaknesses of Baldrige Award applicants demonstrate the importance of purchasing's strategies and actions in the assessment of supplier quality.

Strengths

- "The adoption of 'customer concept,' with the supplier as the customer of purchasing, is supported by positive actions."
- "The approach to supplier quality as a continuation of the applicant's TQM philosophy is supported by supplier training and supplier involvement in the design process. Suppliers provide SPC data."
- "Supplier certification is used as a supplier rating system."
- "Supplier agreements cover 70 percent of the total dollars spent in the applicant's purchases, and the applicant plans to extend supplier agreements to all suppliers."

- "Principal quality requirements are specified and include an acceptance rate of 99.99 percent and on-time deliveries of 95 percent."
- "End users of suppliers' products are surveyed annually to determine degree of satisfaction with the suppliers."
- "Key suppliers are audited semiannually."
- "Suppliers that receive highest levels of certification are publicly recognized by the applicant in leading business publications and newspapers."
- "Applicant gives assistance, including training, to suppliers to help them meet and maintain standards."
- "Supplier representatives serve in applicant's quality improvement teams."
- "Applicant has a strategy to reduce the number of suppliers."

Weaknesses

- "It is unclear how specific quality requirements are defined and communicated to suppliers."
- "A supplier-recognition system implemented in 1900 appears to be too new and too soon to have produced results."
- "There is little evidence of any effort to reduce supplier lead times."
- "Provisions of the supplier agreements seem to be reactive rather than proactive, emphasizing inspection and repair service rather than prevention."
- "It appears that the extent of cooperation to form the relationship is limited."
- "The measurement of quality of the annual supplier product quality survey is not described."
- "The extent of direct measurement of supplier quality levels is unclear."
- "The supplier audit process is not defined in sufficient detail to be evaluated."

	Categories	Items	Areas to Address	Points
Total exam	7	28	89	1000
Purchasing influence	4	8	12	225
		Purchasing		
	2.0 Information and analysis	2.2 Competitive comparisons and bench-marks	2.2.b(4) "Supplier performance"	25
	3.0 Strategic qual-ity planning	3.1 Strategic qual-ity and com-pany perfor-mance plan-ing process	3.1.a(5) "Supplier capabilities"	35
		3.2 Quality and performance plans	3.2.b(1) "Key perform-ance indicators deployed to… suppliers"	25
	5.0 Management of process quality	5.1 Design and in-troduction of quality prod-ucts and services	5.1.b(3) "Supplier capability"	40
		5.5 Quality assessment	5.5.b "Supplier requirements"	15
		5.3 Process man-agement: bus-iness processes and support services	5.3.a In notes section purchasing is classified as a support service	30
		5.4 Supplier qual-ity	5.4.a/b/c All areas focus on suppliers	20
	6.0 Quality and operational results	6.4 Supplier qual-ity results	6.4.a/b All areas focus on suppliers	35

Figure C-1. Purchasing's influence on the Malcolm Baldrige National Quality Award.

5.4 Supplier Quality
(20 pts.)

Describe how the quality of materials, components, and services furnished by other businesses is assured and continuously improved.

AREAS TO ADDRESS
a. approaches used to define and communicate the company's quality requirements to suppliers. Include: (1) the principal quality requirements for key suppliers; and (2) the principal indicators the company uses to communicate and monitor supplier quality.
b. methods used to assure that the company's quality requirements are met by suppliers. Describe how the company's overall performance data are analyzed and relevant information fed back to suppliers.
c. current strategies and actions to improve the quality and responsiveness (delivery time) of suppliers. These may include: partnerships, training, incentives and recognition, and supplier selection.

Notes:

(1) The term "supplier" as used here refers to other-company providers of goods and services. The use of these goods and services may occur at any stage in the production, delivery, and use of the company's products and services. Thus, suppliers include businesses such as distributors, dealers, and franchises as well as those that provide materials and components.

(2) Methods may include audits, process reviews, receiving inspection, certification, testing, and rating systems.

Figure C-2. Baldrige Examination Item 5.4: Supplier Quality. (*1992 Baldrige Application Guidelines*)

6.4 Supplier Quality Results *(35 pts.)*

Summarize trends in quality and current quality levels of suppliers; compare the company's supplier quality with that of competitors and with key benchmarks.

AREAS TO ADDRESS
a. trends and current levels for the most important indicators of supplier quality.
b. comparison of the company's supplier quality levels with those of competitors and/or with benchmarks. Such comparisons could be industry averages, industry leaders, principal competitors in the company's key markets, and appropriate benchmarks. Describe the basis for comparisons.

Note: *The results reported in Item 6.4 derive from quality improvement activities described in Item 5.4. Results should be broken down by major groupings of suppliers and reported using the principal quality indicators described in Item 5.4.*

Figure C-3. Baldrige Examination Item 6.4: Supplier Quality Results. (*1992 Baldrige Application Guidelines*)

Item 6.4 Supplier Quality Results

The guidelines are to "Summarize trends and levels in quality of suppliers and services furnished by other companies." (See Figure C-3.)

Experts' evaluations of various applicants' strengths and weaknesses have included these assessments of purchasing strategies and actions in the area of supplier quality results:

Strengths

- "Applicant has developed partnership relationships with transportation carriers to be more responsive to customer needs."

- "Data suggests that applicant's transportation carriers are best-in-class."
- "Supplier rating system requirements sets the benchmark for suppliers. A goal has been set to use only 100 percent TQM suppliers by 1990."
- "Results from supplier surveys indicate that supplier quality has improved substantially since 1990 and that the improvement is sustained."

Weaknesses

- "There appears to be limited data on supplier performance."
- "Applicant's comparison to world-class suppliers is weak. The comparison is based on the supplier rating system, yet little evidence was provided."
- "The data presented is for a subset of applicant's suppliers."
- "The scope of the measurement presented is limited."
- "It is not clear to what extent the annual supplier questionnaire is a direct measure of supplier quality."

Hints for Applicants

Having had the opportunity to experience the same type of training that Baldrige examiners receive and the opportunity to serve as an ITT internal examiner, I would offer the following hints to those applying:

- Include a thesis statement for each of the 28 items. This will provide a good road map for the reader.
- Address all 89 specific areas. As an examiner, I deducted from an applicant's total score for each area omitted.
- Minimize the risk of miscommunication. Spell out all comments clearly or risk misinterpretation by at least some of the examiners.
- Clearly mark each area being addressed [e.g., 5.7.a (1)]. I found myself inclined to penalize an applicant (even though I wasn't

supposed to) if I had to hunt and interpret comments on areas that were not clearly marked.

- If you disagree with the inclusion of a specific area, indicate the reason and describe your own approach. This is better than not answering.

- State exactly what is being done at your company. Examiners are instructed to "believe" what they read, but remember that unsupported assertions (without plausible data, information, and facts) may not be accepted as valid.

- Be aware that examiners will indicate those areas of concern to be reviewed at a site visit.

A Comparison of the Baldrige Award and the Deming Prize

"Comparing the Baldrige Award to the Deming Prize," said Jack Rubin, vice president of operations at ITT Avionics," is like comparing the Emmy Award to the Oscar."

The Deming Application Prize (known as the Deming Prize) was instituted in 1951 by the Union of Japanese Scientists and Engineers (JUSE) in recognition and appreciation of the contributions made to Japanese industry by Dr. W. Edwards Deming. Although the Baldrige Award has the same general purpose as the Deming Prize—that of recognizing and honoring quality in industry—there are a number of significant differences between the two.

- The Deming Prize is not a contest for winners and losers. It is awarded to all companies that apply, provided they meet the requirements. In the case of applicants who do not meet the standards, the examination process is extended—up to two times in three years. The Baldrige Award is only awarded to the top companies in each of the three categories.

- Compared to the criteria for the Baldrige Award, the Deming criteria are few in number and average only six to seven words in length. It is hard to compare the 23 pages of criteria for the

Baldrige Award to the one-page checklist of criteria for the Deming Prize. The looseness of the Deming guidelines allows the judges to reevaluate their meaning and importance over time without yearly revisions. The Deming checklist has ten major categories:

1. Policy and objectives
2. Organization and its operation
3. Education and its extension
4. Assembling and disseminating information
5. Analysis
6. Standardization
7. Control
8. Quality assurance
9. Effects
10. Future plans

- While the Baldrige Award restricts applications to U.S. companies, the Deming Prize has been open to overseas companies since 1984. (This was a result of strong international interest.) Florida Power and Light is the only U.S. company that has so far won the Deming Prize.
- The cycle for the Baldrige Award is six months from time of application, as compared to one year for the Deming Prize.
- The Deming Prize criteria consistently emphasize the use of statistical methods, while the criteria for the Baldrige Award contain very few mentions of specific statistical techniques.

Summary

Purchasing and the Malcolm Baldrige Award impact each other. The primary use of the award includes the assessment of companies and the assessment of suppliers.

Appendix D
ITT Defense's Supplier Readiness Assessment

Introduction

"Today's subcontractor delivery promises inevitably turn into tomorrow's missed deliveries." If this lament has an all-too-familiar ring to it, take heart; there is a better way.

Experience has convinced us that, as responsible buyers, we must put much more of a broad-based effort into pre-award reviews with our major subcontractors. It is no longer acceptable to conduct only cost and technical reviews before awarding a contract. Today's environment mandates a full-capability review conducted by senior members of the buyer's functional disciplines. That review should ensure that the prospective supplier has

1. The knowledge base to deliver a quality product (the right people)
2. A proven, repeatable process (written procedures/processes)
3. The appropriate resources to produce at maximum rate (tools, test equipment, shop labor)

4. The managerial skills to recognize in-process perturbations in time to effect corrective action

We have found that most subcontractor problems are caused by parts that will meet specification but cannot be built at the required delivery rate. This inability to build to rate may be caused by any one or a combination of the following reasons:

- Lack of or poor design rules
- Poor management of the transition between the model shop prototype and full production
- No formal program of risk identification and risk management
- Lack of production engineering involvement in the design

The following pages describe a method used to manage major subcontracts in a very proactive manner. This method was developed by Michael E. Buck and Frank Colantuono of ITT.

Using the Production Readiness Review

At ITT Defense, we do not wait for a missed delivery to trigger a review of a major subcontractor. Instead, reviews are conducted in a pre-award mode to minimize the potential for a missed delivery.

Major subcontract production readiness reviews are convened by the director of procurement and supported by the product assurance, manufacturing, engineering, operations planning and control, and procurement functions. The review is conducted by panels representing these disciplines in accordance with the following established format:

Day One: Introduction and Overview (8 A.M. to 9:30 A.M.)

1. "Buyer" team chief states the objectives of the review to the "Seller" team.
2. Seller executive presents the seller organization/corporate structure and discusses relationship to parent company, if applicable.

3. Seller team chief presents the program being reviewed and the relationship between R&D efforts and production requirements.

4. Seller team chief presents product assessment and qualification status.

5. Seller team chief presents program status, problems, and risk assessment, together with management plan to minimize risks.

Day One: Panel Discussions Begin (9:30 A.M. to 6 P.M.)

A series of presentations, discussions, procedure reviews, and implementation audits are held between senior members of the buyer's functional disciplines and corresponding functional experts from the seller's prospective manufacturing facility. The buyer's representatives must be completely familiar with the personnel, procedures, processes, and management systems the seller will employ to execute the contract. Panel discussions are carried out in tandem with one another for the sake of expediency.

Day Two: Panel Discussions End/Debriefing (8 A.M. to 12 noon)

Panel discussions are completed by noon on Day 2. Each panel prepares a debriefing presentation, along with "requests for information/action" for the seller management team.

The debriefing presentations cover the general areas of product assurance, production engineering, management and controls, and materials and subcontracts.

The following specific items are addressed during the panel reviews:

Product Assurance: Management structure and effectiveness
                                    Customer interface/relations
                                    Delivery procedures
                                    Quality documentation (records)
                                    Quality assurance plans

	Interchangeability
	Inspection interface
	Assembly quality assurance
	Corrective action
	Material review board
	Process control
	Risk assessment (identification and reduction)
Production Engineering:	Facility requirement (current/future)
	Fabrication production (method/risk)
	Product flow
	Assembly concepts
	Tooling concepts
	Planning paper (fabrication/assembly)
	Producibility plans
	Corrective action systems
	Test planning
	Risk assessment (identification and reduction)
Management and control:	Load/capacity plans
	Inventory management
	Configuration management
	Performance measurement
	Manpower planning (fabrication/assembly)
	Budget planning
	Lot size analysis
	Production control staffing/plan
	Make/buy analysis
	Schedule coordination
	Systems application (production management)
	Risk assessment (identification and reduction)
Material and subcontracts:	Procurement plans
	Purchase order control

Supplier schedule control
Interfaces (program office, product
assurance, manufacturing, etc.)
Purchased part management
Customer furnished material
Supplier labor contract status
Risk assessment (identification and
reduction)

Panel members rate each vendor in accordance with the following code for its respective area:

Rating Code	Rating Description
Blue	Exceeds expectations.
Green	All systems, procedures, and implementation in good order. No action required of this vendor.
Yellow	Some corrections are required by the vendor.
Red	Supplier does not meet minimum requirement; significant improvement is required prior to contract award.

The key success factor in this review is selecting the correct individuals from the buyer's line management cadre. Knowledgeable, hands-on personnel are required to conduct this review, since the buyer must ascertain not only that a procedure exists but that the procedure is adequate and has been implemented as written. This part of the evaluation/review must be conducted in the appropriate workplace. If the task is to evaluate the adequacy of the vendor's system for dealing with discrepant materials, the reviewer must conduct an on-site evaluation of the physical area which the vendor allocates to quarantine discrepant material. This area should be monitored for limited accessibility (employees should not have open access to this area). The contents of the area should be evaluated for aging of materials retained in it. (Lengthy storages indicate the lack of a system to deal with such material or the lack of adherence to the vendor's procedures.)

If an evaluator elected to evaluate scheduling methodology, a review of the policy or procedure would only be a starting point. Analysis of the supplier's schedule contained in the manufacturing plan would have the same value as reading the procedure. (Anyone can make milestones line up to appear that an on-time delivery will occur.)

Using the Schedule Technique Evaluation Process

The schedule evaluator must pick a currently-in-production program or contract and conduct an actual time verification. This evaluation may be performed equally well on a vendor that has a totally automated scheduling system or a manual system. Any scheduling system must be checked for validity. The following procedure should be implemented to conduct the schedule verification effort:

1. List the supplier's current committed schedule to the buyer.
2. List the supplier's internal target schedule in support of Step 1.
3. Establish product flow through the supplier's factory; flowchart with times supplier uses to plan product.
4. Determine dedicated, versus shared, production resources.
5. Obtain copies of the current shop schedule.
6. Determine whether the supplier's internal schedule supports the supplier's current commitment by comparing Steps 1 and 2.
7. Using data accumulated in Steps 2 and 3, determine whether lower-level schedules support internal target schedules. Trace several legs of the product flow from the deliverable-item level through the entire process, documenting cumulative assembly/subassembly/component requirements at each level in the product structure. The following procedure may be implemented to determine which month on the internal target schedule to select as the starting point:

   ▪ Determine the cumulative cycle time for the leg of the product schedule to be audited.
   ▪ Using the cumulative cycle determined, count the same number of months/weeks into the future. For example, if

the cumulative cycle time is twelve weeks, use the end of the current week as week one, and count out twelve weeks into the future to determine the starting point.

8. Using the cumulative minimum requirements developed in Step 7, check the actual detailed schedules to which the vendor is working to verify that detailed schedules support supplier internal targets.

9. Repeat Steps 7 and 8, using supplier's committed schedule instead of internal target schedule.

10. Using planning cycle times identified in Step 3, check the most recent month's work for actual completions, using shop travelers, to determine the appropriateness of planned cycle times. Actual times are working days/weeks between the initial date of the traveler and the completion date of the work order.

11. Visit several work centers on the shop floor and monitor a sizable random sample of work orders on the shop floor to determine overdue work; record specific findings.

12. Using detailed shop schedules obtained in Step 5, verify that those documents are in the manufacturing work centers and ask sufficient questions of the area manufacturing supervisor to establish whether he or she is sufficiently familiar with the documents to indicate actual usage.

The schedule-technique evaluation process is mandatory. Within eight hours, the evaluator will be able to determine whether his product will flow through the vendor's plant.

While this review procedure is both time-consuming and costly, the time and cost of the review would be greatly exceeded by the cost of even one unreliable supplier and the resulting disruption in the buyer's manufacturing process.

Appendix E

Strategies for Doing Business with Small Disadvantaged Businesses

Introduction

As our society changes, small disadvantaged businesses (SDBs) are receiving more and more attention. This appendix explores various strategies for doing business with SDBs. While the discussion focuses chiefly on the concerns of prime contractors to the Department of Defense, the strategies involved are generally applicable to any large company attempting to cultivate and/or improve its business dealings with SDBs.

Background

The focus on small businesses began after World War II with the enactment of the Reconstruction Finance Act. The intent was to recognize and cultivate the innovations and expertise that come from small business enterprises. The contribution of small businesses to the free enterprise system was apparent during World War II, and Congress has continued to support these businesses for nearly 50 years through a variety of legislation.

Of particular importance to Department of Defense (DoD) contractors was Public Law 95–507, enacted in 1979. This law, requiring contractors to submit subcontracting plans for all proposals and subsequent contracts over $500,000 was intended to quantify the efforts made by large companies to work with small and/or small disadvantaged businesses. The legislation and regulations asociated with this program are voluminous, beginning with a series of definitions as to what actually constitutes a small and/or small disadvanteged business. (For the purposes of this discussion, an SDB is a small business concern owned and controlled by socially and economically disadvantaged individuals. This means a 51-or-more percent ownership by one or more black Americans, Hispanic Americans, Native Americans, or Asia-Pacific Americans.) Other regulations cover set-aside programs (in which the government can authorize a contractor to award a contract without bidding) and various reporting requirements for prime contractors.

In 1987, Congress enacted Public Law 99–661, Section 1207, which requires a minimum of a 5 percent goal in the subcontracting plan for awards to SDBs. The intent was to secure a "serious" effort from contractors to place business with SDBs. A further outgrowth of this legislation is Public Law 100–656, which imposes liquidated damages on contractors who fail to achieve their SDB goal as well as their small business goal, which it sets at 20 percent. The intent, again, is to get the contractors to make a serious effort with SDBs and small businesses (SBs). As of October 1, 1991, no contractor has been held liable for liquidated damages, but no company wants to be the first to be prosecuted.

Old Strategy

For many years, the efforts of the SB/SDB program were largely centered on the procurement area. This focus, however, limited the impact in dollar achievement for SDBs. The items for which procurement has unlimited authority over source selection are normally generic items that tend to have low dollar value. The high-dollar items tend to be those defined by engineering specifications or original-equipment-manufacturer (OEM) capital equipment items. Engineering specification items are usually limited by specific source control designations. This means that the manufacturer is specified, or is limited by prior qualification testing costs, which tend to make it cost-prohibitive to second or third source an item. Distributors of electronic and electro-mechanical components are common SDB sources for procurements but, as previously stated, the dollar amount for this type of purchase is usually limited.

Procurement professionals tried to expand their sources by attending procurement fairs and using directories. The fairs yielded limited results because the majority of attendees at such fairs are supply companies: component distributors, computer suppliers, industrial suppliers, etc. The directories often proved to be time wasters because so many of them fail to screen the companies they list or even update their files for each new edition to weed out companies no longer in existence.

Those companies that channeled their SDB program through procurement usually had very limited success. Prime contractors in this category showed achievement below 2 percent, while the industry average was in the 3 percent range. The pressures exerted by the regulations forced all prime contractors to reexamine their SDB programs and refocus where necessary.

New Strategy

The current task for prime contractors is how to set the appropriate goals and then meet them while still following such basic TQM philosophies as reducing the supplier base. The strategies of TQM that call for overall company involvement are applicable to the SDB program activities. Those prime contractors that have been reasonably successful in their SDB program use activities that involve all company functions and originate from top management.

In order of priority, the key elements in a successful SB/SDB program are:

1. The recognition and public endorsement of the program by top management

2. An annual budget to support the required activities
3. A company-wide awareness of the program and its requirements
4. Formal procedures that promote, direct, and control the consideration of SDBs in the design or development of the program

Top Management Support

Experience indicates that the recognition and public endorsement of top management can give an SDB program the impetus it needs to be successfuly executed. By demonstrating the company's commitment to the program, visible management support encourages employees to be more responsive in their respective roles. (Management's involvement, however, must be seen to be genuine, not merely a temporary lip service paid to the goals of the program.) Methods of communicating the support of top management might include letters to the employees from the company president endorsing the SDB program, and/or offical publications, disseminated to the employees, publicizing the program's policies, procedures, and management support.

Annual Budget

An annual budget is obviously necessary to pursue such SDB program activities as trade fairs, seminars, and meetings; advertisements; donations of surplus equipment; developmental assistance; and training. There are definite costs associated with the execution of an SDB program, and the existence of a formal budget to cover them is further evidence that the company is serious about the program.

The extent of cost commitment may vary widely and still be effective at either extreme. Specifically, the cost effectiveness of public relations activities that demonstrate a good-faith effort is not measurable with any degree of accuracy.

The elements of a good-faith effort can range from advertising to attract potential SDB suppliers, to all-day outings for potential SDB suppliers. Obviously, the cost impact will vary accordingly.

An example of a reasonably cost-effective activity is an in-house seminar for selected SDBs, sponsored and hosted by the prime contractor. Such a seminar might include a breakfast, a presentation about doing business with the prime contractor, a plant tour, and meetings with individual buyers. The annual budget must be able to cover the costs involved in providing such developmental assistance to SDBs and other minimal outreach activities.

Company-Wide Awareness

Ongoing program success will depend on a company-wide awareness program. Employees must be aware that there is an SDB program in place in order to participate in it. Specifically, those employees involved in source selection—i. e., engineers, requisitioners of capital equipment, and plant engineering personnel—must automatically consider SDBs in their source selection process. Also, those employees involved in donations of surplus equipment must routinely consider historically black colleges and universities (HBCU) and minority institutions (MI) in their disposal procedure. The prime contractor receives credit for such gifts at the donation value approved by IRS practices and shows the donation as an achievement on the SF295 quarterly report to the government.

Formal Company Procedures

Formal company procedures are absolutely necessary to insure the success of an SDB program. All disciplines should routinely consider SDBs in their source selection process, and their direction and control should be routinely documented. Documented procedures might include company methods for considering HBCUs and MIs as potential recipients of donated surplus equipment; SDB awareness training in the company's overall TQM training program; and provision for advertising and mailing costs in the company marketing budget.

The time it takes for any SDB program to succeed depends in part on the business base of the prime contractor, but with appropriate groundwork, ultimate success is generally assured. If, due to an insufficient business base, contract purchases are not available for placement with SDBs, high-dollar indirect purchases can be explored. SDB sources for high-dollar indirect items include travel agencies, security services, maintenance services, and office supply companies. One of the keys to success is to eliminate the stereotypes associated with SDBs. Another is to use the SDBs' disadvantaged status as a banner. Suppliers should always be sought for their expertise first, and only then for their other characteristics—i. e., for their status as small or small disadvantaged businesses.

Mentor/Protege Program

SDBs with special expertise sometimes need developmental assistance from a prime contractor. When this happens, the prime contractor can be reimbursed through the Pilot Mentor/Protege Program, under which the prime contractor, or mentor, receives reimbursement (in the form of money or credit toward the contractor's overall SDB achievement) for costs incurred in assisting the SDB, or protege. The intent of the program is to provide prime contractors with an incentive to help SDBs develop—for the good of both the SDBs and the marketplace.

TQM must be considered in this effort. The declining supplier base makes it all the more imperative that the SDBs raise their performance level. We cannot achieve Total Quality Management in purchasing if we limit our supplier base to large businesses and established small businesses only. If an SDB needs assistance in some part of its operation—in its quality assurance system, its manufacturing division, or even its personnel department—it is incumbent upon the prime contractor to provide it. The benefit will be experienced in the development of a long-term relationship that is advantageous to both parties.

Appendix F
Purchasing
TQM-isms

Purchasing's role in TQM will involve behaviors that may not be initially accepted by the organization. Listed below are twenty-one of these truths:

1. "It's OK to make a mistake."
2. "Everything is not going great."
3. "TQM takes time."
4. "The use of teams for continuous improvement will take longer than the manager directing the solution expects."
5. "If it doesn't get measured, it doesn't get done."
6. "It may be necessary to put a measurement chart on almost everything that moves."
7. "TQM is not measured by the size of layoffs."
8. "One of purchasing's customers is the supplier."
9. "Most of the time the supplier is not wrong."
10. "It's acceptable for suppliers to make a profit."
11. "Pay bills on time and take discounts only when deserved."
12. "Solve problems by being like Detective Colombo (pay attention to details and dig until the root cause is found), not by

emulating Superman and using a lot of power to end a crisis quickly."

13. "Flavor-of-the-month programs will melt away by the heat of the summer."

14. "Technology techniques, such as SPC, Kanban, and Taguchi, may not be the answer."

15. "TQM is not just for the troops."

16. "Buyers must practice MBWA: Management By Wandering Around."

17. "Videos and slogans may not work—you need eye-to-eye contact."

18. "Middle management may stall, more than lead, the TQM effort."

19. "TQM requires a *commitment* from management, not just involvement."

20. "Quality does not lie outside the realm of people's real jobs."

21. "The SPC Code is: In God we trust, all others bring data."

Bibliography

Books

Ansari, A., and B. Modarress, *Just-In-Time Purchasing*, The Free Press, 1990.

Camp, Robert C., *Benchmarking: The Search for Industry Best Practices That Lead to Superior Performance*, ASQC Quality Press.

Caplan, Frank, *The Quality System*, Chilton Book Co., 1989.

Carlisle, John, and Robert Parker, *Beyond Negotiations*, John Wiley and Sons, Ltd.

Crosby, Phil, *Quality Is Free*, McGraw-Hill, 1979.

Ernst and Young Quality Consulting Group, *Total Quality: An Executive Guide for the 1990s*, Business One, 1990.

Grieco, Peter, Michael Gozzo, and Jerry Claunch, *Just-In-Time Purchasing*, PT Publications, 1989.

Johnson, Ross, and Richard Weber, *Buying Quality*, ASQC Quality Press, 1985.

Kundra, Dennis, *The Purchasing Manager's Decision Handbook*, Cahners Publishing Co., 1975.

Oakland, John, *Total Quality Management*, Heinemann Professional Publishing Ltd.

Scherkenbach, William W., *The Deming Route to Quality and Productivity*, Mercury Press/Fairchild Publications.

Schorr, John, and Thomas Wallace, *High Performance Purchasing*, Oliver Wright Publications, 1986.

Articles

Altany, David, "Share and Share Alike," *Industry Week*, July 1991.

Barclift, Jane, Lt.–USN, "Strategic Partnerships: Competitive Advantage or Risky Business?" *NAPM Insights*, January 1991.

Bemowski, Karen, "The Benchmarking Bandwagon," *Quality Progress,* January 1991.

Bohte, Keki, "Strategic Supply Management," American Management Association, 1989.

Bowles, Jerry, "The Race to Quality Improvement" (advertisement), *Journal for Quality and Participation.*

Bryce, G. Rex, "Quality Management Theories and Their Application," *Quality* magazine, February 1991.

Bush, David, and Kevin Dooley, "The Deming Prize and the Baldrige Award: How They Compare," *Quality Progress,* January 1989.

Carbone, James, "How Purchasing Helped IBM Win the Baldrige," *Electronics Purchasing,* March 1991.

——————, "Raytheon Looks for a Few Good Suppliers," *Electronics Purchasing,* June 1991.

——————, "Purchasers Crack the Quality Whip," *Electronics Purchasing,* December 1990.

——————, "Purchasing Helped Xerox Win the Baldrige," *Electronics Purchasing,* March 1990.

Cavinato, Joseph, "Benchmarking New Ways of Measuring Purchasing," *Purchaser,* December 1990.

Cayer, Shirley, "Welcome to Caterpillar's Quality Institute," *Purchasing* magazine, August 1990.

Cohodas, Marilyn, "Help Your Firm Win the Baldrige in 1990," *Electronics Purchasing,* March 1990.

Cole, Robert F., "Comparing the Baldrige and the Deming," *Journal for Quality and Participation,* July/August 1991.

Conway, John, "Reducing the Cost of Materials," *John Diebold and Associates Brief,* No. 1.1.2.

Conway Quality, Inc., "The Right Way to Manage," 1989.

Coopers and Lybrand, "Supplier Partnership within the Aerospace and Defense Industry," September 1990.

Crawford, C. C., "The Crawford Slip Method," *Air Force Journal of Logistics,* Spring 1985.

D'Alessandro, Jennifer, "Supplier Certification Headaches," *Purchasing Management,* September/October 1990.

Dillon, Thomas, "Work Value System Measures Productivity," *Purchasing World,* 1990.

Ellram, Lisa, "The Supplier Selection Decision in Strategic Partnerships," *Journal of Purchasing and Materials Management,* October 1990.

Evans-Correia, Kate, "How Xerox Made Its Comeback," *Purchasing,* January 1991.

Fargher, John, "C-1 Implementing Total Quality Management," *Aerospace and Defense Symposium Proceedings,* 1990.

Frank, Donald, "Procurement to Time-Phased Requirements: A Challenge for Aerospace and Defense Procurement People," *LA-ADSIG Digest*, Edition VII, January 1990.
Gardner, Frank, "Gauge 90s Suppliers by Their Quality of Service," *Electronics Purchasing*, July 1991.
————, "'New' NCR Buys Technology First," *Electronics Purchasing*, December 1990.
————, "Time to Get More Buying for Your Buck," *Electronics Purchasing*, August 1990.
Geldman, Steven, "Supplier Recognition Award," *Quality*, October 1990.
Glover, M. Katherine, "Malcolm Baldrige National Quality Award: The Quest for Excellence," *Business America*, November 1989.
Goddard, Walter, "The Art of Winning Vendors: Cooperation and Participation," *Purchasing*, January 1986.
Gorny, Traci, "Predicting the 90s in Purchasing," *NAPM Insights*, January 1991.
Gossmann, John, and Judy Murphree, "Matchmaker, Make Me a Match," *NAPM Insights*, February 1990.
Gray, David, and Dennis Berg, "C-4 Supplier Certification through Total Quality Management," *American Production and Inventory Society*. 1990 Aerospace and Defense Symposium Proceedings, 1990.
Harwood, Charles, and Gerald Pieters, "It Pays to Know the Cost of Ownership," *Electronics Purchasing*, May 1989.
Hay, Edward J., "Implementing JIT Purchasing: Phase IV—Relationship Building," *The Magazine of Manufacturing Performance*, April 1990.
Hunter, Nicholas R., "Catalyst for Change," *Purchasing*, November 1990.
Johnson, John, "Successful TQM Is a Question of Leadership," *Journal for Quality and Participation*, June 1990.
Kendrick, John J., "Customers Win the Baldrige Award Selections," *Quality*, January 1991.
Lanza, Julie, "At Your Service, Motorola," *Electronics Purchasing*, July 1991.
Lester, Martin, "Taking Benchmarking a Step Further," *NAPM Insights*, July 1991.
Limperis, Jim, "When Single Source Leads to Remorse," *NAPM Insights*, September 1990.
Lynch, Robert, and William Silvia, "Developing a Strategic Alliance," *NAPM Insights*, June 1991.
Main, Jeremy, "How to Win the Baldrige Award," *Fortune*, April 23, 1990.
McElroy, John, "QFD Building the House of Quality," *Automotive Industries*, January 1989.

Morgan, James, "Comparable Productivity: A Goal for Purchasing," *Purchasing,* March 1991.

Morgan, James, and Susan Zimmerman, "Building World-Class Supplier Relationships," *Purchasing,* August 1990.

Murphree, Julie, "Purchasing's Customers Speak Out," *NAPM Insights,* August 1991.

Musselwhite, Ed, and Linda Moran, "On the Road to Self-Direction," *Journal for Quality and Participation,* June 1991.

O'Halloran, J. David, S. A. Walleck, and Charles Leader, "Benchmarking World-Class Performance," *The McKinsey Quarterly,* Number 1, 1991.

Placek, Chester, "Role of Baldrige Award in Quality Improvement Becoming Subject of Debate," *Quality,* February 1991.

Raia, Ernest, "JIT Delivery: Redefining On-Time," *Purchasing,* September 1990.

————, "1990 Medal of Professional Excellence," *Purchasing,* September 1990.

Relman, Curt W., "The Baldrige Award: Leading the Way in Quality Initiatives," *Quality Progress,* July 1989.

Rice, Trudy Thompson, "Partnering in Purchasing," *NAPM Insights,* February 1990.

Rosen, Carol, "Why IC Buying Will Be a Team Effort in the 90s," *Electronics Purchasing,* January 1991.

Sanger, David, "U.S. Suppliers Get a Toyota Lecture," *New York Times,* October 1990.

Stainbrook, Craig W., "All the Wrong Reasons," *Electronic Buyers News,* October 1990.

Stocker, Gregg, "TQM and the Purchasing Profession," *NAPM Insights,*

Stork, Ken, "A World-Class Customer," *Electronic Buyers News,* July 1990.

Sullivan, Joanne, and Tom Werner, "Ten Power Tools for Facilitators," *Journal for Quality and Participation,* December 1990.

Thier, Marian, "Steering the Leadership," *Journal for Quality Participation,* June 1990.

Thor, Carl G., "Fundamental Performance Measurements," *NAPM Insights,* October 1990.

Tonkin, Lea, "Changing the Rules in the Supplier-Buyer Game," *Manufacturing Systems,* October 1989.

Watkins, Luan, "I Am Committed but How Do I Show It?" *Journal for Quality and Participation,* June 1990.

————, "Partnering: The First Step Is to Stop Focusing Only on Cost Issues," *Total Quality Newsletter,* June 1991, Volume 2, Number 6.

Other

Aluminum Company of America, "Benchmarking: An Overview of Alcoa's Benchmarking Process," 1990.

American Production and Inventory Control Society 1989 Conference Proceedings:
  Saathoff, John, "Supplier Partnerships That Create Value."
  Olsen, Robert E., "Vendor Certification: Pathway to an Effective JIT Relationship."

American Productivity and Quality Center (APQC) Benchmarking Clearinghouse Design Steering Committee Meeting, June 18, 1991.

American Productivity and Quality Center (APQC) Current Training Offerings (*Consensus*, Volume 3, Number 2) 1990.

Center for Advanced Purchasing Studies, Arizona State University, Studies on Purchasing Performance Benchmarks, 1991.

Electronics Industries Association, "Report of Winter Meeting for Materials Procurement Committee," Government Division, March 14, 1990.

Department of Commerce, National Institute of Standards and Technology.

Metamorphosis of a Global Competitor, Richard F. Alban, Alban Associates.

Minutes of Symposium and Executive Summary on Supplier Partnerships within the Aerospace and Defense Industries, Coopers and Lybrand Management Consulting Services, 1990.

News Release from News Office, Massachusetts Institute of Technology, "Made in America, a Two-Year, Eight-Industry MIT Study," May 1989.

Procurement Quality Control, edited by James L. Bosset, ASQC Quality Press, 1988.

Report of the Joint OSD-Air Force-Industry Total Quality Management Impediments Process Action Team Findings and Recommendations, June 1989.

Total Quality Management Guide, Department of Defense, DoD 5000.51g-Final Draft 2/15/90.

Total Quality Management at ITT Defense, education module, Revision 1, June 1991.

Xerox Corporation, "Competitive Benchmarking," 1987.

Index

About the Author

James Cali has over 20 years of hands-on purchasing experience in both the commercial and defense sectors. Starting as a buyer at Westinghouse Electric Corp., he progressed to purchasing and materials management positions there. At ITT Avionics in New Jersey, Mr. Cali served as manager of TQM and planning, manager of operations strategy, and manager of procurement. He is currently Director of Materials at Harvard Industries/Elastic Stop Nut Division. His experience also includes service as CEO of a small manufacturing company, which gave him unique insight into the assistance needed by the purchasing executive from a CEO for implementation of purchasing programs. A graduate of the U.S. Military Academy at West Point, Cali holds an MBA from Mount St. Mary's College. He is an active member and speaker of APICS, NAPM, and ASQC.